KEY NOTES ON AGRICULTURAL ECONOMICS, BUSINESS MANAGEMENT AND STATISTICS

For Ready Reference to the

STUDENTS, TEACHERS, RESEARCHERS & ASPIRANTS OF COMPETITIVE EXAMINATIONS

THE EDITORS

Dr. U.D. Chavan obtained his M.Sc. (Agri. in Biochemistry) degree from Mahatma Phule Krishi Vidyapeeth, Rahuri. He received his Ph.D. degree in Food Science from Memorial University of Newfoundland St. John's Canada in 1999. He has done International Training on "Global Nutrition 2002" at Uppsala University Uppasala, Sweden in 2002. Dr. Chavan worked as Senior Research Assistant in the Department of Biochemistry & Food Science and Technology at MPKV Rahuri from 1988 to 2000. During his Ph.D., he worked as Technician/Research Associate at Atlantic Cool Climate Crop Research Center and Agriculture and Agri-Food Canada. He received D.Sc. degree in 2006 from USA.

Dr. Chavan is presently working as a Senior Cereal Food Technologist in the Department of Food Science & Technology at Mahatma Phule Krishi Vidyapeeth, Rahuri.

Dr. J.V. Patil obtained his M.Sc. (Agri.) from, MPKV, Rahuri. He completed his course work for Ph.D. at CCSHAU, Hisar and research at MPKV, Rahuri in 1992. He rendered his research and teaching services at MPKV Rahuri as Geneticist, Associate Professor, Plant Breeder and Professor of Genetics & Plant Breeding and Head, Genetics and Plant Breeding Department, MPKV, Rahuri. He also delivered many administrative responsibilities in the University. Dr. Patil joined as the Director, Directorate of Sorghum Research, Hyderabad in August 2010.

THE CONTRIBUTORS

Dr. P.P. Pawar is an Assistant Professor in the Department of Agriculture Economics at Mahatma Phule Krishi Vidyapeeth, Rahuri.

Dr. H.R. Shinde is a Senior Research Assistant in the Department of Agriculture Economics at Mahatma Phule Krishi Vidyapeeth, Rahuri.

Dr. D.S. Nawadkar is an Assistant Professor in the Department of Agriculture Economics at Mahatma Phule Krishi Vidyapeeth, Rahuri.

Dr. C.A. Nimbalkar is an Associate Professor in the Department of Statistics at Mahatma Phule Krishi Vidyapeeth, Rahuri.

KEY NOTES ON AGRICULTURAL ECONOMICS, BUSINESS MANAGEMENT AND STATISTICS

For Ready Reference to the

STUDENTS, TEACHERS, RESEARCHERS & ASPIRANTS OF COMPETITIVE
EXAMINATIONS

Editors

U.D. CHAVAN
&
J.V. PATIL

Contributors

P.P. PAWAR
H.R. SHINDE
D.S. NAWADKAR
C.A. NIMBALKAR

2015
Daya Publishing House®
A Division of
Astral International (P) Ltd
New Delhi 110 002

Published by	:	**Daya Publishing House®**
		A Division of
		Astral International Pvt. Ltd.
		– ISO 9001:2008 Certified Company –
		4760-61/23, Ansari Road, Darya Ganj
		New Delhi-110 002
		Ph. 011-43549197, 23278134
		E-mail: info@astralint.com
		Website: www.astralint.com
Laser Typesetting	:	**Twinkle Graphics, Delhi**
Printed at	:	**Thomson Press India Limited**

PRINTED IN INDIA

PREFACE

India is an agricultural country. The Indian economy is basically agarian. Inspite of economic and industrialization, agriculture is the backbone of the Indian economy. As Mahatma Gandhi said "India's lives in villages and agriculture is the soul of Indian economy". Agriculture is a vast subject and encompasses at least 20 major and minor subjects in it. New developments have lead to entirely a new face of agriculture. Study of agriculture has always been intrigued with a mosaic of interwove concepts, subjects, facts and figures. There are number of books and large literature on Agricultural Economics, Business Management and Statistics but the Key Notes type of book have not been compiled in a readable manner.

The present book *"Key Notes on Agricultural Economics, Business Management and Statistics"* has been designed to fulfill this long felt need of students, teachers, researchers and aspirants of competitive examinations. It is designed in such a way that give rapid, easy access to the core materials in a short format which facilitates easily learning and rapid revision. The book carries fundamentals of Agricultural Economics, Business Management and Statistics. The book is divided in two parts. The Part A of the book is Key Notes on Agricultural Economics and Business Management and the Part B of the book is on Key Notes on Statistics. The most recent information is provided along with a detailed list of references for further reading.

Hope this book would be highly useful for graduate and post-graduate students of agriculture, teachers and researchers. This book will also useful for the aspirants of various competitive examinations such as Agricultural Research Service (ARS), ICAR- National Eligibility Test (NET), State Eligibility Test (SET), Junior Research Fellowship (JRF), Senior Research Fellowship (SRF), Civil Services, Allied Agricultural Examinations and Extension Workers for reference and easy answers of many complicated questions. Thus it is expected that this book will adequately meet the need of wider circle of students and readers for preparing their professional career.

We acknowledge the references that are used in this manuscript. Authors are also thankful to all scientists and friends who have helped directly or indirectly while preparing this manuscript. The editors of grateful to all the contributors for their cooperation, support and timely submission of their manuscripts for bringing out this publication. We would have like to acknowledge the patience and support of our families whilst we have spent many hours with drafts of manuscripts rather than with them. Lastly, our sincere thanks to publisher Astral International Pvt. Ltd., New Delhi who provides an opportunity to publish this book.

To all readers we extend an invitation to report that no doubts have escaped our attention and to offer suggestion for improvements that can be incorporated in future editions.

U.D.Chavan and J.V. Patil

Editors

CONTENTS

KEY NOTES ON AGRICULTURAL ECONOMICS AND BUSINESS MANAGEMENT

1

DISCOVERIES

Name of Scientist	Law / Principle / Contribution
Adam Smith	'An enquiry into the nature and causes of wealth which treats of wealth' emphasis to wealth. Subsistence theory of wages
Safffle in Germany	Placed the role of man in economics higher than that of wealth and Droz in France
Marshall	Wealth and man equal importance, Utility analysis-Diminishing *marginal utility theory, Law of demand D=F(P) Introduced concept of consumer's surplus
Robbins	"Nature and Significance of Economic Science" Economics is not concerned with desirability
J.M. Keynes	General theory of employment and income
R.G.D. Allen and J.R. Hicks	Indifference curve-ordinal measurement of utility. IC is popular and essential paraphernalia of modern economics
Prof Samuelson	Revealed Preference theory, Fundamental Theorem of consumption theory
Hicks	Revised theory of demand- Logical weak ordering theory
Neumann-Morgenstern	Statistical utility theory
Armstrong	Marginal preference theory
Pareto Vilfred	Welfare–Any change harms no one and make people some better off
Kaldor and Hicks	Welfare–Gainer at higher level than looser
Scitovsky	Welfare–Double Criterian
Bergson	Formulation of set of explicit value judgment
N. Senior, J.S. Mill	Economics should concern with What is and not what ought to be Economics is positive. Deductive method for economic enquiry simple generalization to particular

Name of Scientist	Law / Principle / Contribution
K.K. Dewett	Economics cannot be disassociated from ethics. Economics is both positive and normative
Historical school - Roscher, Hildebrand, Fedric L. Cliff Leslie	Inductive/realistic method for economics—particular to general
John Stuart Mill	First use concept of Statics and Dynamics in Economics
Robert Giffen	Giffen Paradox–demand is strengthened with price or weakened with fall in price
Thomas Robert Malthus	Essay on principles of Population Theory
David Ricardo	Theory of Rent
J.B. Clerk	Development of marginal productivity theory
J.S. Mill	Wage fund theory
Karl Marx	Theory of exploitation - wages
Walker	Residual claimant theory of wages (Output remained after rent, interest and profit goes to worker)
Taussig	Modified marginal productivity theory of wages
John Rae, Bohm	Theories of interest -1. Productivity (demand),
Bawerk Keynes, Hicks -Hansen	2. Abstinence/waiting (supply), 3. Austrian/Agio, 4. Fisher time preference, 5.Liquidity Preference, 6.Classical/real 7. Modern

THE SVERIGES RIKSBANK PRIZE IN ECONOMIC SCIENCES IN MEMORY OF ALFRED NOBEL FROM 1969-2010

Name of the Scientist(s)	Year	Contribution for which Nobel prize is given
Ragnar Frisch Jan Tinbergen	1969	"for having developed and applied dynamic models for the analysis of economic processes"
Paul A. Samuelson	1970	"for the scientific work through which he has developed static and dynamic economic theory and actively contributed to raising the level of analysis in economic science"
Simon Kuznets	1971	"for his empirically founded interpretation of economic growth which has led to new and deepened insight into the economic and social structure and process of development"
John R. Hicks Kenneth J. Arrow	1972	"for their pioneering contributions to general economic equilibrium theory and welfare theory"
Wassily Leontief	1973	"for the development of the input-output method and for its application to important economic problems"
Gunnar Myrdal Friedrch August von Hayek	1974	"for their pioneering work in the theory of money and economic fluctuations and for their penetrating analysis of the interdependence of economic, social and institutional phenomena"
Leonid Vitaliyevich Kantorovich Tjalling C. Koopmans	1975	"for their contributions to the theory of optimum allocation of resources"
Milton Friedman	1976	"for his achievements in the fields of consumption analysis, monetary history and theory and for his demonstration of the complexity of stabilization policy"
Bertil Ohlin James E. Meade	1977	"for their path breaking contribution to the theory of international trade and international capital movements"
Herbert A. Simon	1978	"for his pioneering research into the decision-making process within economic organizations"
Theodore W. Schultz Sir Arthur Lewis	1979	"for their pioneering research into economic development research with particular consideration of the problems of developing countries"
Lawrence R. Klein	1980	"for the creation of econometric models and the application to the analysis of economic fluctuations and economic policies"
James Tobin	1981	"for his analysis of financial markets and their relations to expenditure decisions, employment, production and prices"
George J. Stigler	1982	"for his seminal studies of industrial structures, functioning of markets and causes and effects of public regulation"
Gerard Debreu	1983	"for having incorporated new analytical methods into economic theory and for his rigorous reformulation of the theory of general equilibrium"
Richard Stone	1984	"for having made fundamental contributions to the development of systems of national accounts and hence greatly improved the basis for empirical economic analysis"
Franco Modigliani	1985	"for his pioneering analyses of saving and of financial markets"
James M. Buchanan Jr.	1986	"for his development of the contractual and constitutional bases for the theory of economic and political decision-making".

Contd...

Name of the Scientist(s)	Year	Contribution for which Nobel prize is given
Robert M. Solow	1987	"for his contributions to the theory of economic growth"
Maurice Allais	1988	"for his pioneering contributions to the theory of markets and efficient utilization of resources"
Trygve Haavelmo	1989	"for his clarification of the probability theory foundations of econometrics and his analyses of simultaneous economic structures"
Harry M.Markowitz, Merton H. Miller, William F. Sharpe	1990	"for their pioneering work in the theory of financial economics"
Ronald H. Coase	1991	"for his discovery and clarification of the significance of transaction costs and property rights for the institutional structure and functioning of the economy"
Gary S. Becker	1992	"for having extended the domain of microeconomic analysis to a wide range of human behaviour and interaction, including nonmarket behaviour"
Robert W. Fogel Douglass C. North	1993	"for having renewed research in economic history by applying economic theory and quantitative methods in order to explain economic and institutional change"
John C. Harsanyi cooperative John F. Nash Jr. Reinhard Selten	1994	"for their pioneering analysis of equilibria in the theory of non-games"
Robert E. Lucas Jr.	1995	"for having developed and applied the hypothesis of rational expectations, and thereby having transformed macroeconomic analysis and deepened our understanding of economic policy"
James A. Mirrlees William Vickrey	1996	"for their fundamental contributions to the economic theory of incentives under asymmetric information"
Robert C. Merton Myron S. Scholes	1997	"for a new method to determine the value of derivatives"
Amartya Sen	1998	"for his contributions to welfare economics"
Robert A. Mundell	1999	"for his analysis of monetary and fiscal policy under different exchange rate regimes and his analysis of optimum currency areas"
James J. Heckman Daniel L. McFadden	2000	"for his development of theory and methods for analyzing selective samples" and "for his development of theory and methods for analyzing discrete choice"
George A. Akerlof A. Michael Spence Joseph E. Stiglitz	2001	"for their analyses of markets with asymmetric information"
Daniel Kahneman Vernon L. Smith	2002	"for having integrated insights from psychological research into economic science, especially concerning human judgment and decision-making under uncertainty" and "for having established laboratory experiments as a tool in empirical economic analysis, especially in the study of alternative market mechanisms"
Robert F. Engle III Clive W. J. Granger	2003	"for methods of analyzing economic time series with time-varying volatility (ARCH)" and "for methods of analyzing economic time series with common trends (cointegration)"
Finn E. Kydland	2004	"for their contributions to dynamic macroeconomics: the time consistency of economic policy and the driving forces behind business cycles"

Contd...

Name of the Scientist(s)	Year	Contribution for which Nobel prize is given
Robert J. Aumann Thomas C. Schelling	2005	"for having enhanced our understanding of conflict and cooperation through game-theory analysis"
Edmund S. Phelps	2006	"for his analysis of intertemporal tradeoffs in macroeconomic policy"
Leonid Hurwicz Eric S. Maskin Roger B. Myerson	2007	"for having laid the foundations of mechanism design theory"
Elinor Ostrom Paul Krugman	2008	"for his analysis of trade patterns and location of economic activity"
Elinor OStrom Oliver E. Williamson	2009	"for her analysis of economic governance, especially the commons"
Peter A Diamond Dale T. Mortensen Christopher A Pissarides	2010	"For their analysis of markets with search frictions"

NOBEL LAUREATES IN ECONOMICS

Year	Nobel Laureates in Economics - Contribution
2008	Paul Krugman -Analysis of trade patterns and location of economic activity
2007	Leonid Hurwicz, Eric S. Maskin, Roger B. Myerson-Laid foundations of mechanism design theory
2006	Edmund S. Phelps — Analysis of inter-temporal tradeoffs in macroeconomic policy
2005	Robert J. Aumann, Thomas C. Schelling- Understanding of conflict and cooperation through game-theory analysis
2004	Finn E. Kydland, Edward C. Prescott - contributions to dynamic macroeconomics: the time consistency of economic policy and the driving forces behind business cycles"
2003	Robert F. Engle III for methods of analyzing economic time series with time-varying volatility (ARCH)"
	Clive W.J. Granger - for methods of analyzing economic time series with common trends (cointegration)
2002	Daniel Kahneman-human judgment and decision-making under uncertainty Vernon L. Smith - the study of alternative market mechanisms
2001	George A. Akerlof, A. Michael Spence, Joseph E. Stiglitz- analyses of markets with asymmetric information

Contd...

Year	Nobel Laureates in Economics - Contribution
2000	James J. Heckman - development of theory and methods for analyzing selective sample , Daniel L. McFadden-development of theory and methods for analyzing discrete choice
1999	Robert A. Mundell - analysis of monetary and fiscal policy under different exchange rate regimes and his analysis of optimum currency areas
1998	Amartya Sen-contributions to welfare economics
1997	Robert C. Merton, Myron S. Scholes - a new method to determine the value of derivatives
1996	James A. Mirrlees, William Vickrey - contributions to the economic theory of incentives under asymmetric information
1995	Robert E. Lucas Jr. -developed and applied the hypothesis of rational expectations, and thereby having transformed macroeconomic analysis and deepened our understanding of economic policy
1994	John C. Harsanyi, John F. Nash Jr., Reinhard Selten - pioneering analysis of equilibria in the theory of non-cooperative games
1993	Robert W. Fogel, Douglass C. North-renewed research in economic history by applying economic theory and quantitative methods in order to explain economic and institutional change
1992	Gary S. Becker - extended the domain of microeconomic analysis to a wide range of human behaviour and interaction, including nonmarket behaviour
1991	Ronald H. Coase - discovery and clarification of the significance of transaction costs and property rights for the institutional structure and functioning of economy
1990	Harry M. Markowitz, Merton H. Miller, William F. Sharpe - pioneering work in the theory of financial economics
1989	Trygve Haavelmo - clarification of the probability theory foundations of econometrics and his analyses of simultaneous economic structures
1988	Maurice Allais - pioneering contributions to the theory of markets and efficient utilization of resources

Contd...

Year	Nobel Laureates in Economics - Contribution
1987	Robert M. Solow - contributions to the theory of economic growth
1986	James M. Buchanan Jr. - development of the contractual and constitutional bases for the theory of economic and political decision-making
1985	Franco Modigliani - pioneering analyses of saving and of financial markets
1984	Richard Stone - fundamental contributions to the development of systems of national accounts and hence greatly improved the basis for empirical economic analysis
1983	Gerard Debreu - incorporated new analytical methods into economic theory and for his rigorous reformulation of the theory of general equilibrium
1982	George J. Stigler - seminal studies of industrial structures, functioning of markets and causes and effects of public regulation
1981	James Tobin - analysis of financial markets and their relations to expenditure decisions, employment, production and prices
1980	Lawrence R. Klein - creation of econometric models and the application to the analysis of economic fluctuations and economic policies
1979	Theodore W. Schultz, Sir Arthur Lewis - pioneering research into economic development research with particular consideration of the problems of developing countries
1978	Herbert A. Simon - pioneering research into the decision-making process within economic organizations
1977	Bertil Ohlin, James E. Meade - path breaking contribution to the theory of international trade and international capital movements
1976	Milton Friedman - consumption analysis, monetary history and theory and for his demonstration of the complexity of stabilization policy
1975	Leonid Vitaliyevich Kantorovich, Tjalling C. Koopmans - contributions to the theory of optimum allocation of resources

Contd...

Year	*Nobel Laureates in Economics - Contribution*
1974	Gunnar Myrdal, Friedrich August von Hayek - theory of money and economic fluctuations and for their penetrating analysis of the interdependence of economic, social and institutional phenomena
1973	Wassily Leontief - development of the input-output method and for its application to important economic problems
1972	John R. Hicks, Kenneth J. Arrow - contributions to general economic equilibrium theory and welfare theory
1971	Simon Kuznets - founded interpretation of economic growth which has led to new and deepened insight into the economic and social structure and process of development
1970	Paul A. Samuelson - developed static and dynamic economic theory and actively contributed to raising the level of analysis in economic science
1969	Ragnar Frisch, Jan Tinbergen - developed and applied dynamic models for the analysis of economic processes

2

ABBREVIATIONS

Abbreviation	Full form
AAP	Agricultural and Allied Products
AAP	Annual Action Plan
ACD	Agricultural Credit Department
ACIAR	Australian Centre for International Agricultural Research Asian Development Bank
ACs&ABCs	Agri Clinics and Agri Business Centers
ADB	Agricultural Development Branch
AEZ	Agri-Export Zone
ADB	African Development Bank
AFC	Agricultural Finance Corporation
AFMA	Association of Food Marketing Agencies.
APO	Asian Productivity Organization
AICC	All India Congress Committee
AIGBWO	All India Gramin Bank Workers' Organization
AIRCRC	All India Rural Credit Review Committee
AIRDISC	All India Rural Debt and Investment Survey Committee
AOA	Agreement on Agriculture
AP	Achievers' Resources Private Limited
APC	Armored personnel carrier
APC	Agricultural Prices Commission
APCCADB	Andhra Pradesh Central Co-operative Agricultural Development Bank
APCOBARD	Andhra Pradesh Co-operative Bank for Agriculture and Rural Development

Abbreviation	Full form
APEC	Asia pacific economic cooperation
APEDA	Agricultural and Processed Food Products Export Development Authority.
APFAGMS	Assistance to States for Infrastructure Development of Export Agricultural Technology Management Agency
BAIF	Bharatiya Agro Industries Foundation
APL	Above Poverty Line
ATREEATIC	Ashoka Trust for Research on Environment and Ecology Agricultural Technology and Information Centre
AWSR	Automatic Weather Stations
BMGFBMI	Bill and Melinda Gates Foundation Body Mass Index
ARDC	Agricultural Refinance and Development Corporation
ASAC	Asian Standards Advisory Committee
ASI	Annual Survey of Industries/Agricultural Situation in India
ASE	Bovine Spongiform Encephalopathy
ATM	Automatic trained machine
AUD	Ambedkar University, Delhi
AVORD	Association of Voluntary Organizations for Rural Development.
BCM	Community Forest Management
BCR	Benefit-Cost Ratio
BE	Budget Estimates
BICP	Bureau of Industrial Cloths and Prices
BIRD	Bankers Institute for Rural Development
BIS	Bureau of Indian Standards
BLBC	Block Level Bankers Committee
BOP	Balance of Payments

Abbreviation	Full form
BoT	Balance of Trade
BPL	Below Poverty Line
BRIMSS	Block Rural Industries Marketing and Servicing Society
CAB	College of Agricultural Banking
CACP	Commission for Agricultural Costs and Prices
CADA	Command Area Development Authority
CALCOB	Committee on Agricultural Loans through Commercial Banks
CAM	Centre for Agricultural Marketing
CAPS	Cover and Plinth Storage.
CAS	Credit Authorization Scheme
CCA	Capital Consumption Allowance
CCB	Central Cooperative Bank
CCEA	Cabinet committee on economic affairs
CCI	Cotton Corporation of India
CCPA	Cabinet committee on political affairs
CERC	Consumer Education and Research Centre.
CFB	Corrugated Fiber Board
CGPCS	Contact group on piracy off the coast Somalia
CGTMSE	Credit guarantee fund trust for micro and small enterprises
CIFP	Cost, Insurance and Freight Price.
CIFE	Central Institute of Fisheries Education
CII	Confederation of Indian Industries
CIP	Central Issue Price
CIS	Crop Insurance Scheme
CIAH	Central Institute for Arid Horticulture
CICEF	Central Institute of Coastal Engineering for Fishery
CIDA	Canadian International Development Agency
CIFA	Central Institute of Freshwater Aquaculture

Abbreviation	Full form
CIFNET	Central Institute of Fisheries Nautical and Engineering Training
CIFRI	Central Inland Fisheries Research Institute
CIFT	Central Institute of Fisheries Technology
CMFP	Comprehensive Marine Fisheries Policy
CMFRI	Central Marine Fisheries Research Institute
CMIE	Centre for Monitoring Indian Economy.
CMZ	Coastal Management Zone
CSAMB	Council of State Agricultural Marketing Boards
CPI	Consumer Price Index.
CPSE	Central public sector enterprise
CRAFICARD	Committee to Review Arrangements for Institutional Credit for Agriculture and Rural Development
CRR	Cash Reserve Ratio
CRZ	Coastal Regulation Zone
CSC	Central Seeds Committee.
CSCB	Central Seed Certification Board.
CSCs	Common Service Centers
CSE	Center for Science and Environment
CSIR	Council of Scientific and Industrial Research
CSO	Central Statistical Organization.
CSOs	Civil Society Organizations
CSR	Certified Shareholder Report
CSR	Corporate Social Responsibility
CSS	Centrally Sponsored Schemes
CSWRCTI	Central Soil and Water Research and Conservation Training Institute
CU	Carcass Unit
cu km	cubic kilometer
cum	cubic meter

Abbreviation	*Full form*
CWC	Central Warehousing Corporation
CWC	Central Water Commission
DAC	Department of Agriculture and Cooperation
DAHD	Department of Animal Husbandry and Dairying
DIDA	Danish International Development Agency
DBOD	Department of Banking Operations and Development
DCC	District Consultative Committee
DCCB	District Central Co-operative Bank
DCP	District Credit Plan
DES	Directorate of Economics and Statistics.
DFID	Department for International Development
DICGC	Deposit Insurance and Credit Guarantee Corporation
DCBARD	District Co-operative Bank for Agriculture and Rural Development
DIT	Department of Information Technology
DJRC	D.J. Research & Consultancy
DMI	Directorate of Marketing and Inspection
DoAC	Department of Agriculture and Cooperation
DoAH	Department of Animal Husbandry
DoF	Department of Fisheries
DoH	Department of Horticulture
DPIP	District Poverty Initiative Project
DRDA	District Rural Development Agency
DTC	Direct tax code
EAS	Electronic Auctioning System.
EDA	International Development Association
EC	European Commission
ECAFE	Economic Commission for Asia and Far East
EDI	Economic Development Institute

Abbreviation	Full form
EEIs	Extension Education Institutes
EEZ	Exclusive Economic Zone
EFL	Ecologically fragile lands
EG	Eastern Ghats
EM	Emerging market
EOU	Export Oriented Unit.
EPCG	Export Promotion Capital Goods
EPTD	Environment and Production Technology Division
EPW	Economic and Political Weekly
ERM	Extension, Rehabilitation and Modernization
EXIM	Export-Import.
F&V	Fruits and Vegetables
FAO	Food and Agriculture Organization
FAQ	Fair Average Quality.
FCI	Food Corporation of India/Fertilizer Corporation of India
FCI	Food Corporation of India
FDA	Food and drug administration
FDA	Fish Farmers Development Agencies/Farmers Field School
FGP	Food-grain Production
FGP	Food-grain Production
FHP	Farm Harvest Price
FICCI	Federation of Indian Chamber of Commerce and Industry
FIVIMS	Food Insecurity and Vulnerability Information and Mapping System
FOBP	Free on Board Price
FORP	Free on Rail Price
FPR	Flood Prone River
FPS	Fair price Shops

Abbreviation	Full form
FSI	Forest Survey of India
FSM	Food Security Mission
FSSA	Food Safety and Standards Act
FSSAI	Food Safety and Standards Authority of India
FTZ	Free Trade Zone
FY	Financial Year
FYP	Five Year Plan
GAFTA	Grain and Feed Trade Association
GAP	Good Agricultural Practices
GATT	General agreement on tariffs and trade
GCC	Girijan Cooperative Cooperation
GCF	Gross Capital Formation
GDP	Gross domestic products
GEF	Global Environment Facility
GHP	Good Hygiene Practices
GIS	Geographic Information System
GJEPC	Gems and jewellery export promotion council
GLV	Green Leafy Vegetables
GM	Genetically Modified
GNP	Gross National Product
GoD	Government of Denmark
GoJ	Government of Japan
GoI	Government of India
GONL	Georgia Organization of Nurse Leaders
GW	Ground Water
GoUK	Georgia Organization of Nurse Leaders Government of United Kingdom
ha	Hectare
HACCP	Hazard Analysis and Critical Control Point
HPCMPS	Horticultural Producers Cooperatives Marketing and Processing Society

Abbreviation	Full form
HP	Harvest Price (Harvest Season Price)
HP	Himachal Pradesh
HPMC	Himachal Pradesh Horticultural Processing and Marketing Corporation
HRD	Human Resource Development
HSRP	High security registration plate
HTM	Horticulture Technology Mission (for the Northeast and other Hill States)
HYV	High Yielding Variety
IARI	Indian Agriculture Research Institute
IBF	India business forum
IBRD	International Bank for Reconstruction and Development
ICAR	Indian Council of Agricultural Research
ICC	Indian Chamber of Commerce
ICFRE	Indian Council for Forest Research and Education
ICM	Integrated Crop Management
ICMR	Indian Council of Medical Research
ICRIER	Indian Council for Research on International Economic Relations
ICRISAT	International Crop Research Institute for Semi Arid Tropics
ICT	Information and Communication Technology
ID	Identity Document
IDA	International Development Agency
IDP	Internally displaced production
IFAD	International Fund for Agricultural Development
IFP	Integrated Fisheries Project
IFPRI	International Food Policy Research Institute
IGSI	Indian Grain Storage Research Institute
IIASA	International Institute for Applied System Analysis

Abbreviation	Full form
IIFT	Indian Institute of Foreign Trade
IIM	Indian Institutes of Management
IIP	Index for Industrial production
IIP	Indian Institute of Packaging
IIT	Indian Institutes of Technology
IJAM	Indian Journal of Agricultural Marketing
ILO	International Labour Organization
ILRI	International Livestock Research Institute
IMD	Indian Meteorological Department
IMF	International Monetary Fund
INCOIS	Indian national center for ocean information services
INM	Integrated Nutrient Management
INR	Indian Rupees
INRM	Integrated Natural Resource Management
IONS	Indian ocean naval symposium
IOSC	International organization of securities commission
IPM	Integrated Pest management
IPR	Intellectual Property Rights
IRDA	Insurance regulatory & development authority
IRRI	International Rice Research Institute
ISAM	Indian Society of Agricultural Marketing
ISI	Indian Standards Institution
ISMR	Indian Summer Monsoon Rainfall
ISOPOM	Integrated Scheme of Oilseeds, Pulses, Oil Palm and Maize
IT	Information Technology
ITTO	International Tropical Timber Organization
IWMI	International Water Management Institute
J&K	Jammu and Kashmir
JBIC	Japanese Bank for International Cooperation

Abbreviation	Full form
JCI	Jute Corporation of India
JFM	Joint Forest Management
JLG	Joint Liability Group
JMDC	Jute Manufactures Development Council
JPC	Joint Parliamentary Committee
JTM	Jute Technology Mission
KCC	Kisan Credit Card
KHDP	Kerala Horticulture Development Programme
KIAM	Karnataka Institute of Agricultural Marketing
km	Kilometre
KUMS	Krishi Upaj Mandi Samiti
KVIC	Khadi and Village Industries Commission
KVKs	Krishi Vigyan Kendras
LAB	Local Area Bank
LCIA	London court of international arbitration
LFA	Logical Framework Approach
LOOP	Law of One Price
LOP	Letter of Permit
MAP	Medicinal and Aromatic Plants
MBV	Monodon Baculo Virus
MCs	Market Committees
MDB	Multilateral Development Banks
MDG	Millennium Development Goals
MEP	Minimum Export Price
MFI	Micro Financial Institution
MFRA	Marine Fishing Regulation Act
MGREGA	Mahatma Gandhi Rural Employment Guarantee Act
MI	Minor Irrigation
MIC	Market Information Centers

Abbreviation	Full form
MIS	Market Intervention Scheme
MIS	Management Information System
MIT	Minor Irrigation Tank
mm	millimeter
MMA	Macro Management Scheme
MMI	Major and Medium Irrigation
MMTC	Minerals and Metals Trading Corporation
MNC	Multi National Corporation
MoA	Ministry of Agriculture
MoEF	Ministry of Environment and Forests
MoFPI	Ministry of Food Processing Industries
MoRD	Ministry of Rural Development
MOU	Memo of understanding
MoU	Memorandum of Understanding
MoWR	Ministry of Water Resources
MP	Madhya Pradesh
MPC	Marginal Propensity to Consume
MPDC	Market Planning and Design Centre.
MPEDA	Marine Products Export Development Authority
MRTP	Monopolies and Restrictive Trade Practices.
MRV	Measurable Reportable and Verifiable
MS	Marketable/Marketed Surplus
MSP	Minimum Support Price
MSSRF	M.S. Swaminathan Research Foundation, Chennai
MTA	Mid-Term Appraisal
NAARM	National Academy of Agricultural Research Management
NAAS	National Academy of Agricultural Sciences.
NABARD	National Bank for Agricultural and Rural Development.
NAEDB	National Afforestation and Eco-Development Board

Abbreviation	Full form
NACMFD	National Agricultural Cooperative Marketing Federation of India
NAIP	National Agricultural Innovation Project
NAMA	Non-Agricultural market access
NAPCC	National action plan on climate change
NARP	National Agricultural Research Project
NARS	National Agricultural Research System
NATP	National Agricultural Technology Project
NBFGR	National Bureau of Fish Genetic Resources
NCAEPR	National Centre for Agricultural Economics and Policy Research
NCDC	National Co-operative Development Corporation.
NCF	National Commission on Farmers
NCHRH	National council for human resource in Health
NCIWRDP	National Commission for Integrated Water Resource Development Plan
NCPCR	National commission for protection of child rights
NDC	National Development Council
NDDB	National Dairy Development Board
NE	North Eastern
NEC	North Eastern Council
NEH	North Eastern Hills
NELP	New exploration and licensing policy
NER	North Eastern Region
NERAMC	North Eastern Regional Agricultural Marketing Corporation
NFDB	National Fisheries Development Board
NFWF	National Fish Workers Forum
NFHS	National Family Health Survey
NFSM	National Food Security Mission
NFZ	No fire zone

Abbreviation	Full form
NGO	Non-Government Organization
NHB	National Horticulture Board
NHM	National Horticulture Mission
NIAM	National Institute of Agricultural Marketing.
NICT	National Informatics Centre Network
NIN	National Institute for Nutrition
NIOH	National Institute for Occupational Health
NMTPF	National Medium Term Priority Framework
NNMB	National Nutrition Monitoring Bureau
NPCBB	National Project for Cattle and Buffalo Breeding
NPP	Nitrogen Phosphate and Potash
NPW	Net Present Worth
NR	Natural Resources
NRAA	National Rainfed Area Authority
NRCCWF	National Research Centre on Coldwater Fisheries
NREGA	National Rural Employment Guarantee Act
NREGS	National Rural Employment Guarantee Scheme
NREP	National Rural Employment Programme
NRHM	National Rural Health Mission
NRM	Natural Resource Management
NSC	National Seed Corporation
NSDP	National State Domestic Product
NSS	National Sample Survey
NSSO	National Sample Survey Organization
NTFP	Non-Timber Forest Produce
NW	North Western
NWDA	National Water Development Agency
NWDPRA	National Water Development Programme for Rainfed Area
NWMA	National Watershed Management Agency

Abbreviation	Full form
NWP	National Water Policy
O&M	Operation and Maintenance
OBC	Other Backward Classes
OECD	Organization for Economic Cooperation and Development
OGL	Open General License
PAC	Public accounts committee
PAC	Primary Agricultural Cooperative
PACCS	Primary Agricultural Credit Cooperative Societies
PBM	Parity Bound Models.
PDPP	Prevention of damage to public property
PDS	Public Distribution System
PEM	Protein Energy Malnutrition
PFDCs	Precision Farming Development Centers
PHM	Post Harvest Management
PHT	Post Harvest Technology.
PIA	Project Implementation Agency
PIC	Prior Informed Consent
PIM	Participatory Irrigation Management
POP	Persistent Organic Pollutant
PPLPI	Pro-Poor Livestock Policy Initiative
ppm	parts per million
PPP	Public-Private Partnership
PADA	Professional Association for Development Action
PRF	Portfolio Risk Fund
PRI	Panchayati Raj Institutions
QRs	Quantitative Restrictions.
R&D	Research and Development
RSCMF	Rajasthan State Co-operative Marketing Federation
RBI'	Reserve Bank of India

Abbreviation	Full form
RCCS	Rural Credit Card Scheme
RCDF	Rajasthan Co-operative Dairy Federation.
REDD	Reducing Emissions from Deforestation and Degradation
REGP	Rural Employment Generation Programme
RH	Relative Humidity
RIDF	Rural Infrastructure Development Fund
RKVY	Rashtriya Krishi Vikas Yojana
RML	Royal market light
RPDS	Revamped Public Distribution System.
RRB	Regional Rural Bank
RSAMB	Rajasthan State Agricultural Marketing Board
RSWC	Rajasthan State Warehousing Corporation
RVP	River Valley Projects
SAARC	South Asian Association for Regional Cooperation
SAMB	State Agricultural Marketing Board
SAMETI	State Agricultural Management and Extension Training Institute
SSI	Small Scale Industries
SAU	State Agriculture University
SC	Scheduled Castes
SCB	State Cooperative Bank
SDoA	State Department of Agriculture
SEZ	Special Economic Zone
SFAC	Small Farmers' Agribusiness Consortium
SFHE	Small Farmers Horticultural Estates
SFR	State Forest Research
SGSY	Swarnajayanti Gram Swarozgar Yojana
SHG	Self Help Group
SIDBI	Small Industries Development Bank of India
SIRD	State Institutes for Rural Development

Abbreviation	Full form
SMP	Statutory Minimum Price
SMS	Small message service
SPF	Specific Pathogen Free
SPM	Sanitary and Phytosanitary Measures
SREP	Strategic Research Extension Plan
SRI	Systematic Rice Intensification
SRR	Seed Replacement Rate
SRT	Sir Ratan Tata Trust
ST	Scheduled Tribes
STC	State Trading Corporation
SW	Surface Water
SWAN	State Wide Area Network
SWC	State Warehousing Corporation
TBT	Technical Barriers to Trade
TCP	Technical Cooperation Programme
TERI	The Energy Research Institute
TIFAC	Technology Information, Forecasting and Assessment Council
TM	Terminal Markets
TMC	Technology Mission on Cotton
ToT	Terms of Trade
TPDS	Targeted Public Distribution System
TRIFED	Tribal Cooperative Marketing Development Federation of India Limited
TRIPR	Trade Related Intellectual Property Rights
TSI	Technical Support Institute
UIDAI	Unique identification authority of India
UMPP	Ultra mega power project
UNCCD	United Nations Convention to Combat Desertification
UNDAF	United Nations Development Assistance Framework

Abbreviation	Full form
UNDP	United Nations Development Programme
UNEP	United Nations Environment Programme
UNFCCC	United Nations Framework Convention on Climate Change
UP	Uttar Pradesh
USAID	United States Agency for International Development
USDA	United States Department of Agriculture
UTs	Union Territories
VAD	Vitamin A Deficiency
VAT	Value Added Tax
VCRC	Vector Control Research Centre
VFPC	Vegetable and Fruit Promotion Council
WB	World Bank
WBCIS	Weather Based Crop Insurance Scheme
WCP	Women Component Plan
WFP	World Food Programme
WG	Western Ghats
WPI	Wholesale Price Index
WSSV	White Spot Syndrome Virus
WTO	World Trade Organization
WUA	Water Users' Association
WWF	World Wide Fund

3

BASIC CONCEPTS IN AGRICULTURAL ECONOMICS

1. ECONOMICS: MEANING, DEFINITION SCOPE AND IMPORTANCE

According to Adam Smith : Economics was concerned with, 'An enquiry into the nature and causes of wealth which treats of wealth'. In this definition a key position was assigned to wealth. The emphasis shifted from wealth to man, in 19th century - humanistic character of economics comes to be well recognized. Safffle in Germany and Droz in France placed the role of man in economics higher than that of wealth. Man occupied primary place and wealth secondary. Marshall puts, 'Economics is on one side a study of wealth and other and more important side a part of the study of man'.

Robbins in "Nature and Significance of Economic Science" stressed non-material services, e.g., doctor, lawyer etc. They are scarce and have value which has given them their status as economic goods. He defined, 'Economics is the science which studies human behaviors as a relationship between ends and scarce means which have alternatives'.

Individual choices having no social implications, so cannot matter of economics. The theory of economic growth or economic development has not considered. It does not explain problem of unemployment and abundance of manpower rather than scarcity in some countries. The human touch is missing entirely. Robbins has made economics more abstract and complex and hence difficult and unfruitful. This detracts from its utility for the common man.

The credit for bringing about a revolution in economic thinking goes to late lord J.M. Keynes. According to him, 'Economics studies how the levels of income and employment in a community are determined'. In Keynesian terms, Economics is defined as, 'The study of administration of scarce resources and of the determinants of income and employment'. In other wards, it studies the causes of economic fluctuations to see how economic stability could be promoted. In short, Economics may be defined as a 'social science concerned with the proper uses and allocation of resources for the achievement and maintenance of growth with stability'.

SCOPE OF ECONOMICS

Scope means the sphere of study. Scope of economics can be brought out by studying subject matter of economics and as economics is social science, Wants- Efforts-Satisfaction of human being in the social set up forms the subject matter. In primitive society hungry man pick up fruit or hunt animal to satisfy himself. In modern society man started producing goods on large scale and started exchange of goods for meeting the needs. The term exchange required selling and buying. Now a days production of goods in factories require four factors-land, labor, capital and management/organization. These four factor are paid rewards-rent, wages, interest and profit, respectively. Economics studies how these incomes are determined and distributed. This process of income distribution has also come in between efforts and satisfaction.

Thus subject matter of economics is Efforts-Consumption-- Satisfaction of wants,

Production–Making efforts to create things or goods

Exchange–Money credit and banking

Distribution–sharing all that produce among factors of production

In addition economics studies public finance, provides such formula as solution for practical problems. Thus economics is both light giving and fruit bearing. Hence, economics is both a science and an art. It explore and explain the facts, similarly it advocates and condemns the facts. It prescribes rule of life and lives judgment as to what is right and what is wrong.

Positive science explains causes and effects of the things. A normative science discusses the rightness or wrongness of the things. Economics tells us how things happen and also good or bad. It tells us about distribution of wealth and also causes of unequal distribution its goodness or otherwise. Therefore economics is both positive and normative sciences.

Economics tells us how a man utilized his limited sources for the satisfaction of his unlimited wants. Both money and goods are required for satisfaction of wants to promote human welfare. This is subject matter of economics.

Economics-social science–Economics studies human being living in organized society exchanging his goods for others. Thus he obtains his needs from others and satisfies needs of others. Economics, thus is a social science studying individuals of the society and society as a whole.

Subject matter of Economics–Micro and Macro economics relates to price and income theory, respectively. Price theory explains composition/allocation/ production level. Income theory explains level of total production and why level rises and falls. Analyzing problems of economy as a whole is macro-economics. Analysis of behavior of particular decision making unit, firm, industry, consumers is microeconomic-unit.

Macro-economics–is concerned with aggregate and average of entire economy viz., national income, aggregate output, total employment, total consumption. It deals with how an economy grows. It analyses the chief determinants of economic development and the various stages and processes of economic growth. Economic growth is a long run problem- Harrod and Dormer extended the Keynesian analysis (short run problem and economic fluxion) to long run problem of growth and stability.

IMPORTANCE OF ECONOMICS

(A) Theoretical-Informative, mental training, understanding economic system

1. Informative-Economics teaches us many interesting and instructive facts about human behaviour when he engaged in economic activity.

2. Mental training : Economic reasoning trains our mind as reasoning.

3. Understanding functioning of economic system. How the complicated system of today functions almost automatically without any control.

4. Teaches mutual dependence, and

5. Useful citizenship.

(B) Practical

1. **Professional value :** The study of economics is very useful in several profession.

2. **Useful for households :** A household will arrange his expenditure much better if he studies economics.

3. **Useful for labor leaders :** To fight the right of labor

4. Solving problem of poverty, employment, food distribution.

5. **Importance for underdeveloped countries :** To remove poverty un-employment and raise the standard of living. Economics guide to economically backward countries.

Importance of micro-economics

1. **Functioning of free enterprise economy :** Allocation of resources

2. How through market mechanism goods and services distributed

3. Determination of relative prices of product and services.

4. The condition of efficiency in production and consumption.

Practically it helps in formulation of economic policies; promote efficiency in production and welfare of masses. Micro-economics tells us how the economy operates - positive science and also how it should be operated to promote general welfare normative Science.

Limitation : It cannot give an idea of economy as a whole, it assumes full employment which is rarer

Importance of Macro-economics : It is helpful in understanding the functioning of economy. It gives a birds eye view of the economic world. It help in formulation of useful economic policies. To regulate aggregate employment and national income, workout national policy. Economic theory seeks to explain fluctuation in national employment and income.

Limitations : Individual is ignored altogether. Individual welfare is aim of economics but increasing saving at the expenses of individual welfare is not wise. The macroeconomics overlooks individual difference- During stable general price level poor individual may suffer increasing food grain prices.

2. BASIC CONCEPTS IN ECONOMICS

Goods : Anything that can satisfy a human want is called goods Goods means commodities we use, they are concrete material and tangible, e.g., land, house, furniture etc.

Services : Are the work, they are not tangible eg. advice of doctor, work done by labour, railway etc.

Kinds of goods : There are seven categories of good and two types each.

1. **(a) Free goods :** Which exist in plenty and for their use no payment is required eg. air, sunshine. Generally these goods are gift of nature.

   (b) Economic goods : Which are scarce and obtained by paying price. Generally these are manmade things and are limited in nature, e.g., wheat, bread, furniture etc. Economic goods form wealth. Increase in economic goods means increases in wealth. Increase in wealth does not mean increase in economic welfare.

2. **(a) Consumers goods :** Goods used by consumer to satisfy wants directly eg. food clothing etc.

   (b) Producers goods : Goods help us to produce other goods e.g., implements machines. These goods satisfy our wants indirectly because they produce goods which in turn satisfy our wants directly.

3. **(a) Material goods :** These are the concrete and tangible goods land cash etc.

   (b) Non-material goods : These are not concrete and in tangible goods, e.g., services.

4. (a) **Transferable goods :** These are goods of which ownership can be changed. These goods are bodily transferred from one place to other.

   (b) **Non-transferable goods :** These are the goods which cannot be transferred. But the services of such goods can be used by other. Usually these goods are personal qualities like skill, ability, beauty, intelligence.

5. (a) **Personal goods :** These goods refer to personal qualities like skill, ability, beauty, intelligence.

   (b) **Impersonal goods :** These goods are called as external goods.

6. (a) **Private goods :** These are property items of private individuals owned by them.

   (b) **Public goods :** These are goods which are common to all and owned by the society collectively, e.g., Hospital, College, Schools.

7. (a) **Durable :** Book, Cloth, table etc.

   (b) **Non-durable :** also called perishable goods. Food, water, milk etc.

Utility : Want satisfying quality in good. Thus utility means the power of good to satisfy a want. A good/thing may be good or bad, but as it satisfy human want it possesses utility. Some commodities on consumption do not give pleasure still they possesses utility example medicine. Utility gives satisfaction.

Utility is subjective : Utility varies from individual to individual and it also varies from time to time and place to place, e.g., cigarette to non-smoker, warm suit in winter/summer. Thus utility depends on circumstance, so subjective.

1. **Form utility :** By changing the form of an article it attains greater utility, e.g., wood to furniture.

2. **Place utility :** Utility can be created or increased by transporting a good from place of production to consumption example wood from jungle to market.

3. **Time utility :** By storing a commodity and selling it at a time of scarcity, its utility at proper time is enjoyed. This is time utility obtained through storage.

Value : Value of commodity refers to the goods that can be obtained in exchange for it. Fresh air cannot be exchanged for anything therefore its value in economics is zero though it has utility.

The value of commodity thus means the commodities or services that we can get in return for it. Value of commodity is its power of commanding other things in exchange.

Qualification of a commodity to have value : Only economic goods have value in economic sense. It must 1. Possesses utility 2. be scare and 3 be transferable/marketable. All three qualities are essential together, otherwise goods have no value.

Price : Value expressed in terms of money is called price. Exchange of goods for goods called barter in old days. In modern times, price of commodity means its money value, *i.e.,* the price it commands in the market. Thus price expresses value in terms of money.

Wealth : In ordinary language wealth means money. Wealth coveys an idea of prosperity and abundance or richness. In economics in addition to money it include anything which has value. The term wealth is synonymous with economic goods. A good cannot be wealth by itself it is not required by man and it has no utility and if does not satisfy the human want. Anything which has value is called as wealth in economics.

Essential Qualities (Attributes of wealth) : Wealth must have utility, Wealth must be scare, Wealth must be transferable or marketable.

Money and Wealth : As money possesses utility, it being scarce and transferable it is wealth. All money is wealth. But all wealth is not money, because money has got its own form while the wealth is in number of kinds, e.g., water, grains, furniture etc. are different forms of wealth and they are not money.

Wants : Man is bundle of desires. His wants are indefinite food, clothing and shelter are necessaries but modern civilised man struggle for better food, fashionable clothing and comfortable lodging etc. Wants vary from individual to individual. They are dependent upon social and economic position, education, temperament and tastes.

Characteristics of Human Wants

1. Human wants are unlimited.
2. A particular want is satiable.
3. Wants are complementary.
4. Wants are competitive.
5. Wants are both competitive and complementary.
6. Wants has alternative.
7. Wants vary with time place and person.
8. Wants vary in urgency and intensity.
9. Wants multiply with civilisation.
10. ants are recurring in nature.
11. Wants change into habit.
12. Wants are influenced by income, salesmanship and advertisement.

13. Wants are result of customs or convention.

14. Present want are more important than future wants.

Wants are classified as Necessaries, Comforts and Luxuries

1. **Necessaries :** things require for existence, efficiency and convention.

2. **Comforts :** things increase happiness and efficiency by comforts.

3. **Luxuries :** things satisfy superfluous want. Luxury is superfluous consumption. Something we could easily do without luxuries are unnecessary and man can lead a healthy and useful life without luxuries.

Money spend on necessaries give cent per cent compensation or returns, money spent on comforts gives less compensation while money spent on luxuries brings negligible returns or compensation.

For poor luxuries are treated as waste and loss but are advocated as.

- Production of luxuries articles creates employment.

- It help in transfer of money from idle rich to active useful poor.

- It adds to the skill of the workers.

- It acts as a stimuli to new inventions and thus helps for technical and industrial progress.

4

SHORT EXPLANATIONS

1. CONSUMPTION : DEMAND AND SUPPLY

Consumption means the satisfaction of our wants by the use of commodities and services. When we consume a commodity we use its wants satisfying quality. Consumption may be defined as destruction of utility.

Types : (1) Direct consumption : When goods satisfies wants directly, e.g., Water, food, cloth etc.

(2) Indirect or productive consumption : When goods are used for producing other goods which satisfy our wants directly. It is a productive consumption.

Importance : (1) Consumption is beginning as well as end of all economic activity. Want is the beginning of the economic effort. Effort produce commodity. Commodity is finally consumed for satisfaction of want. Thus satisfaction is the end of an economic activity.

(2) Consumption gives push to the production.

(3) The liking and disliking of the consumer induce discovery of new products.

(4) Consumption determines price in the market.

(5) Consumption affects all economic activities, economic progress and finally determine the standard of living of the people.

Standard of living refers to necessaries, comforts and luxuries which a person is accustomed to enjoy. It is quality and quantity of their consumption. The things like food, clothing entertainment are daily requirements, their quantities and quality consumed constitute standard of living. It refers to usual scale of expenditure and attitudes and actions concerned with the use of goods and services.

FACTORS GOVERNING STANDARD OF LIVING ARE

1. **Level of national income and output :** High income high standard of living

2. **Level of productivity :** Higher the productivity higher will be output and so standard of living.

3. **Terms of trade :** Refers to rate of exchange between goods exported and imported, it is ratio between prices of goods exported to the prices of goods imported. If a terms of trade is favourable for a country its standard of living will be higher.

4. **Size of population :** Higher population means lower per capita income and so low standard of living.

5. **Distribution of national income :** Unequal distribution leads to high standard of living for few rich and low standard of living for vast majority.

6. **General price level :** Higher general price level, the standard of living will be lower.

7. **Level of education :** The standard of living is generally higher in educated societies than illiterate societies.

8. For an individual the factors affecting are (1) Income (2) Size of family (3) Family tradition (4) Education tastes and temperament (5) Social customs and Conventions (6) General price level.

Standard of living is reflected in family budget. Family budget is a statement which shows how family income is spent on various items of consumption.

Engel's law of family expenditure—

1. As income increases the percentage expenditure on necessaries of life decreases and vice-versa.

2. Percentage expenditure on luxuries and other cultural and recreational wants increases with an increase in income and decrease when income decreases.

3. For lodging or rent, fuel and light percentage expenditure is invariably the same for all incomes.

4. What so ever the income, percentage expenditure of income on clothing is practically same.

CONSUMER BEHAVIOUR - DEMAND ANALYSIS

There are two techniques of analysis of consumers behaviours (1) Utility analysis (2) The indifference curve technique (Modern Utility analysis)

The law of Diminishing Marginal Utility : It is important law of consumption. It states that as we go on consuming a commodity the satisfaction derived from its successive units goes on decreasing. The more we have a commodity the less we want to have, it is the experience of every consumer that each successive unit yield him less and less satisfaction. In other words at each step, its marginal utility goes on decreasing.

Dr. Marshall states the law as: The additional benefit which a person derives from a given increase of his stock of anything diminished with the growth of the stock that he has

Cake unit	1	2	3	4	5	6	7	8
Utility	12	22	30	36	40	42	42	40
Marginal Utility	12	10	8	6	4	2	0	-2

Total utility goes on increasing upto 6th cake. But marginal utility less and less from 2nd cake. This means total utility of cake increasing at decreasing rate. The marginal utility of 7th is zero and then negative. Thus in consumption of commodity, each successive unit yields less and less marginal utility.

Limitation of law : Dissimilar unit, very small unit quality, for long interval rare collection, abnormal person, change in income, desire for property/valuable goods.

Importance of the law of Diminishing Marginal Utility

1. The law guides the government in deciding policy of taxation.
2. The law is useful in determining prices.
3. The law is used in support of socialism towards equal distribution of wealth.
4. The law guide us in household expenditure.
5. **The law is basis for some economic laws :** The law of demand, the law of substitution, the law of consumers surplus, elasticity of demand etc.

The law of Equi-marginal utility/law of substitution

To satisfy the unlimited wants using limited means it is necessary to pick up the most urgent wants. To device maximum satisfaction with available money, spend such a way that every rupee has greater utility in one commodity than other, till the utilities derived from the last rupee spent in two case are equal. In other words, we substitute some units of a commodity of greater utility for some units of less utility, till the two marginal utilities are equal. This is therefore, called law of substitution or the law of equi-marginal utility. Maximum satisfaction can be derived using law of equi-marginal utility/law of substitution.

The Indifference curve Analysis : The basis of the indifference curve technique is that the consumer has a scale of preference and utilities of satisfaction can be compared as greater or equal. The consumer formulate his scale of preference independently of market price, keeping in view the list of commodities in order to their power to satisfy his wants. From this list one combination of two commodities gives him greater, smaller or equal satisfaction as compared with another combination based on scale of preference. An indifference curve represents a level of satisfaction of the consumer from two commodities. It is drawn on the assumption that for all possible points on an indifference curve the total satisfaction remains constant. Indifference curve represents the various levels of satisfaction. The indifference curve analysis does not rest on the assumption that the marginal utility of money remains constant. The marginal utility of money increases when one buys more with a fall in price and is left with less and less money and vice-versa.

The consumer cannot quantify marginal utility/satisfaction obtained but on scale of preference he can indicate that one combination of two goods gives more or less satisfaction as compared to another combination. He has several combinations of two commodities from which he derives same or equal total satisfaction or combinations are equally preferred or derived by him. A, B, C, D are different combinations of apples and mangos to which consumer is indifferent, joining these points we get a continues curve IC each point on it showing equal satisfaction. This is an indifference curve. The set or indifference curve is called indifference map. The indifference curve technique is very popular with modern economics

Consumers surplus put forth by **Alfred Marshall** explains the benefit which a person derives from purchasing at a lower price, things for which he would rather pay an high price than to go without it may be called as consumers surplus. Consumers surplus is the excess of what we are prepared to pay over what we actually pay for a commodity. It is the difference between what we are prepared to pay and we actually pay.

Many commodities such as salt, postcard, newspaper, a matchbox etc. are very useful but they are very cheap we are therefore prepared to pay much more for them than actually pay. From their purchase, therefore we derive a good deal of surplus or extra satisfaction over and above the price that we pay for them.

Importance : In taxation, increasing prices, planning import, opportunities. (1) If consumers surplus is more in particular commodity Govt. can impose tax on (2) If consumers surplus is very high manufactures or business man increase price of commodity. (3) Consumers surplus depends on total utility whereas price depends on marginal utility. (4) The concept is useful in planning import policy. (5) It indicates the advantages of environment of opportunity.

Demand : Meaning, kinds, demand schedule, demand curve law of demand, factors determining demand.

Bober's defn of demand, 'By demand we mean the various quantities of a given commodity or services which consumer would buy in given market in a given period of time at various prices or various incomes or at various prices of related goods'. Demand is desire backed by willingness and ability to pay for the goods/services.

Kinds of demand : (1) Price demand (2) Income demand (3) Cross demand

(1) **Price demand :** Refers to various quantities of commodity or services that consumer would purchase at a given time in market at various hypothetical prices.

(2) **Income demand :** Refers to various quantities of commodity and services which would be purchased at various levels of income.

(3) **Cross demand :** Means the quantities of goods and services which will be purchased with reference to changes in the prices not of this good but of other related goods.

Demand schedule : The demand schedule is a table or chart which shows quantities of commodity demanded at different prices in a given period of time.

(a) Individual consumers demand schedule

Demand schedule for milk

Sr. No.	1	2	3	4	5	6	7	8
Demand (Qty.) lakh lit /day	8	7	6.5	6	5.5	5	4.5	4
Price (Rs./lit.)	10	12	13	14	15	16	17	18

Practical utility of demand schedule : (1) Forecasting demand and planning production, probable profits can be estimated and (2) To the Government it is useful in estimation of revenue from increase or decrease in taxes.

Demand curve : The data given in Individual demand schedule can be plotted by taking the quantities on OX axis and prices at OY axis. Generally with the decrease in prices the quantity purchased have been increased. Similarly, market curve for demand is obtained.

Why demand curve slopes downwards? This is in accordance with the law of diminishing utility. When the price falls, new purchaser enter the market and old purchaser purchase more. The cumulative effect is an extension of demand resulting into downward slope of demand curve.

Exceptional Demand curves : In the periods of acute shortage of commodity, people have tendency to buy more and more even though prices rise. Under such situation demand curve rise upward. (1) Serious shortage (2) Use of commodity confers distinction (3) in sheer ignorance (4) If the price of necessary goes up.

Law of Demand : "The demand varies inversely (in opposite direction) with price. If the price rises demand contracts if the price falls, demand extends or expands." In other words demand increases with a falling price and decreases with rising price. The law of demand defined as "At any given time the demand for a commodity or service at the prevailing price is greater than it would be at a higher price and less than it would be at a lower price."

Exceptions : If shortage is feared, out of fashion goods, a mark of distinction or honor, ignorance or not aware with price. Increase and decrease as extension and contractions.

Increase demand means more demand at same price or the same quantity demanded at higher price.

Decrease of demand means less demand at the same price or the same quantity at a lower price. Contraction in demand means less demand at higher price.

Factors determining demand

1. **Change in fashion :** Out of fashion leads less demand though low price.

2. **Change in weather :** Demand for woolen cloths only in winter.

3. **Changes in quantity of money in circulation :** Increased purchasing power increase demand.

4. **Change in population :** Size and composition of population — Children baby items, old people medicines.

5. **Change in wealth distribution :** Even distribution of wealth leads more demand.

6. **Changes in real income :** Means things are cheap so that with same money people buy more — more demand for comforts and luxuries.

7. **Changes in habit taste, custom :** Taste for tea change to coffee change demand.

8. **Technical progress :** Inventions and discoveries bring new things in market.

9. **Discovery of cheap substitutes :** Vegetable ghee is cheap substitute to ghee.

10. **Advertisement :** An attractive campaign create new demand.

INTER RELATED DEMANDS

Joint demand–When several things are demanded for a joint purpose, e.g., for tea demand for milk and sugar.

Direct demand : The demand for ultimate object is called direct demand eg. house.

Composite demand : The demand for commodity that can be put to several uses is a composite demand *e.g.* coal used for cooking, heating, steam engine etc.

Elasticity of demand : The law of demand suggest when there is change in price the demand for commodities varies inversely with price. This quantity of demand by virtue of which it changes (increases or decreased) is called elasticity demand. Elasticity means sensitiveness or responsiveness of demand to change in price.

Elastic demand : The demand is elastic when with a small change in price there is great change in demand.

Inelastic demand : When a great change in price causes only slight change in demand, it is inelastic demand.

The elasticity is ratio of percentage change in the quantity demanded to the percentage change in the price.

$$\text{Elasticity of demand} = \frac{\text{Proportionate change in demand}}{\text{Proportionate change in price}}$$

Price elasticity of demand- This elasticity measured responsiveness of potential buyers to change in price. It is ratio of per cent demand to per cent price.

Price elasticity = % change in demand / % change in price

e.g., Price falls from Rs. 500 to 400 (20%) demand increase from 400 to 600 (50%) Ep = 50/20 = 5/2 = 2.5%

Cross elasticity of demand refers to change in demand of a good as a result of change in the price of other good.

$$\text{Cross elasticity} = \frac{\text{\% change in purchases of commodity x}}{\text{\% change in price of commodity y}}$$

This type of elasticity arises in case of interrelated goods such as complementary and substitute goods.

Factors determining the Elasticity of Demand : 1. For necessaries (food, clothing) demand is usually less elastic or comparatively inelastic. 2. For commodities having several uses (coal) demand is elastic. 3. Demand for very high and very low priced commodities is usually inelastic, e.g., salt- low priced, Motorcar high priced. 4. Demand for commodities having substitutes is more elastic.5. For luxuries the demand is more elastic.

Importance

1. **For businessmen :** It guides in fixing prices of goods.
2. **For Government :** For planning on taxation policy.
3. In case of joint products price of each depends on elasticity of demand.
4. **Planning industrial production :** If demand is elastic by slightly reducing prices, sales can be increase, so output.

5. **Paradox of poverty in plenty** : Indian farmers condition.

6. **Determination of wages** : If demand for labour is inelastic the wage rates may arise.

7. **In International trade** : The elastic demand for imported goods and inelastic demand for export product is useful.

Supply

Supply means the quantities that a seller is willing and able to sell at different prices. It is obvious that if the price goes up he will offer more for sale and when price goes down, he will offer less to sell. Supply of any goods may be defined as, 'A schedule of respective quantities of the goods which people are ready to offer for sale at all possible prices'. Like demand, supply is also relative to person, place and time. Supply means the quantity actually offered for sale at a certain price.

Stock means the quantity which can be offered for sale, if the conditions are favorable. Stock constitutes potential supply. In case of perishable products like milk, vegetable, fish the stock and supply are the same.

Supply schedule : It is the table or statement showing the quantities of goods offered for sale at various prices. When the prices falls, less is being offered for sale and vice-versa.

Supply curve : The supply schedule can be represented in the form of curve known as supply curve. The supply curve slopes upwards as we go from left to right. It means as the price rises more is being offered for sale and vice-versa

Law of supply : In given market at any given time, the quantity of any goods which people are ready to offer for sale generally varies directly with the prices'. In other words, other things remaining the same as the price of commodity rises its supply is extended and as the price falls its supply is contracted.

Exceptions to law of supply : (1) Hard needs of money — if the seller is badly in need of money he sell his produce at lower price. (2) Price fall predicted—If further fall in prices is predicted, the seller will sale more even if the price falls.

Extension and increase in supply : Extension of supply means more quantity of good is offered for sale at a higher price. Increase in supply means that either more is offered at the same price or the same quantity is offered at lower price.

Contraction and Decrease in supply : Contraction means that less quantity of goods is offered for sale at lower price. Decrease in supply means that less is offered at the same price or same quantity is offered at higher price.

Elasticity of supply : Supply varies directly with the price. This attribute of supply by virtue of which it extends or contracts with a rise or fall in price is known as elasticity of supply. It refers to the sensitiveness of the supply to changes in prices.

Types of Elasticity of supply

(1) **Inelastic supply :** If there is considerable change in price (rise or fall) it leads only a small change in supply (extension or contraction) e.g. perishable products milk, fruits, vegetables high priced articles like motor car.

(2) **Elastic supply :** If a small change in price (rise or fall) leads to a big change in supply (extension or contraction) the supply is elastic, e.g., clothes.

$$\text{Elasticity of supply} = \frac{(\text{Change in amount supplied / Amount supplied})}{(\text{Change in price /Price})}$$

Causes of change in supply : (1) Natural calamities - Flood drought, earthquakes are bound to affect production, which brings about change in supply.

(2) Technical progress - The very conditions of supply are changed by discoveries or improvement in technique of production.

(3) Change in factor price - It factor of production become cheap the supply will increase.

(4) Transport improvement - Improvement in means of transport enhance supply.

(5) Calamities - War and famine affect supply of goods.

(6) Monopolies - The monopolists may deliberately increase or decrease the supply.

(7) Fiscal policy of Govt. also affect the supply .

Interrelated supply

(1) **Joint supply :** Produced in same process- wheat and straw cotton and cotton seed.

(2) **Composite supply :** Substitute or rival source of supply, e.g., light-electricity/gas/kerosene/candles.

Supply schedule for milk

Sr. No.	1	2	3	4	5	6	7	8
Supply (Qty.) lakh lit /day	4	4.5	5	5.5	6	6.5	7	8
Price (Rs./lit.)	10	12	13	14	15	16	17	18

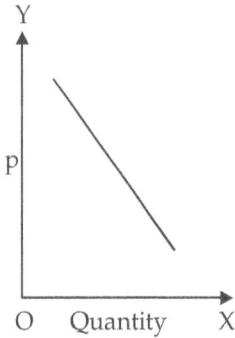
Demand Curve

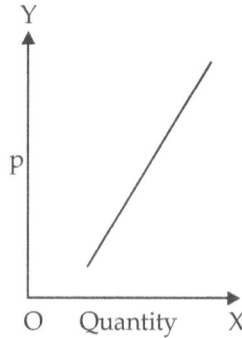
Supply Curve

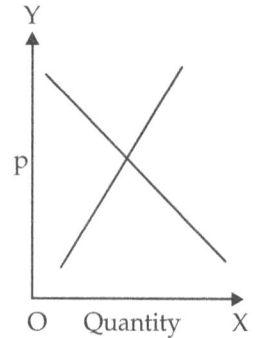
Equilibrium

MARKETS AND PRICE DETERMINATION

Markets for farm products - Meaning and Definition

According to the National commission on Agriculture (XII Report), agricultural marketing is a process which starts with a decision to produce a saleable farm commodity, and it involves all the aspects of market structure, both functional and institutional, based on technical and economic considerations, and include pre- and post-harvest operations, assembling, grading, storage, transportation and distribution.

A study of the agricultural marketing system is necessary

1. For understanding of the complexities involved and

2. Identification of bottlenecks with a view to providing efficient services in the transfer of farm products and inputs from producer to consumer.

The word market comes from the latin word marcatus which means merchandise or trade or a place where business is conducted. Encyclopedia of social sciences (1933) defines market as, 'A market is the area within which the forces of demand and supply converge to establish a single price'. Another definition, given by Gupta (1975), 'Market means a social institution which performs activities and provides facilities for exchanging commodities between buyers and sellers'. Economically interpreted the term market refers not to a place but to a commodity or commodities and buyers and sellers are in free intercourse with one another. Thus, a market is defined in terms of existence of fundamental forces of supply and demand and is not necessarily confined to a particular geographical location.

Essentials/Components of a Market

For market to exist certain conditions must be satisfied. These conditions should be both necessary and sufficient. They are termed as the components of a market.

The existence of a good or commodity for transactions (physical existence is, however, not necessary). The existence of buyers and sellers. Business relationship or inter course between buyers and sellers, Demarcation of area such as place, region, country or whole world.

DIMENSIONS OF A MARKET

There are various dimensions of any specified market these dimensions are :

1.	Location :	6.	Number of commodities
2.	Area or coverage	7.	Degree of competition
3.	Time span	8.	Nature of commodities
4.	Volume of transactions	9.	Stage of marketing
5.	Nature of transaction	10.	Extent of public interaction

Markets may be classified on the basis of each of the above listed dimensional space are described below.

1. **On the basis of location markets are :** Village markets, Primary wholesale markets, Secondary wholesale market, Terminal markets, Seaboard markets.

    (a) *Village markets* : A market which is located in a small village where major transactions take place among the buyers and sellers of a village, is called a village market. It dealt with small quantity of farm produce where the producer sale to consumer example grocery shop or village trader.

    (b) *Primary wholesale markets* : These markets are located in big towns near the centers of production of agricultural commodities. In these markets, a major part of the produce is brought for sale by the producer farmer themselves. Transactions in these markets usually take place between the farmers and traders.

    (c) *Secondary wholesale markets* : These markets are located generally in district head quarters or important trade centers or near railway junctions. The major transactions in commodities take place between the village traders and wholesalers. The bulk of the arrivals in these markets from other markets is handled in large quantities. Therefore, specialized marketing agencies performing different marketing functions, such as those of commission agents, brokers, weigh men, etc. are involved.

    (d) *Terminal markets* : A terminal market is one where the produce is either finally disposed of to the consumer or processors, or assembled for export. Merchants are well organised and use modern methods of marketing. Commodity exchanges exist in these market, which provide facilities for farm forward trading in specific commodities. Such markets are located in metropolitan cities or seaport in Mumbai, Chennai, Kolkata, Delhi.

(e)   *Seaboard markets* : Markets which are located near the seaboard and are meant mainly for the import and/or export of goods are known as seaboard markets ex. Mumbai, Chennai and Kolkata.

2.   **On the basis of area/coverage :** On the basis of area from which buyers and sellers usually came for transactions, markets may be classified into the following four classes—Local or village markets, Regional markets, National markets, World markets.

(a)   *Local or village markets* : a market in which the buying and selling activities are confined among the buyers and sellers drawn from the same village or nearby villages. The village markets exist mostly for perishable commodities in small lots, e.g., local milk market, vegetable market, fruit market etc.

(b)   *Regional markets* : A market in which buyer and sellers for a commodity are drawn from a larger area than the local markets. Regional markets in India usually exist for foodgrains.

(c)   *National markets* : A market in which buyers and sellers are at the national level. National markets are found for durable goods like tea and jute.

(d)   *World market* : A market in which the buyers and sellers are drawn from the whole world. These are the biggest markets from the area point of view. These markets exists in the commodities which have a world wide demand and/or supply, such as coffee, machinery, gold, silver etc. In recent years many countries are moving towards a regime of liberal international trade in agricultural products like raw cotton, sugar, wheat and rice.

3.   **On the basis of time span :** The markets are classified as short period markets, long period markets and secular markets.

(a)   *Short period markets* : The markets which are held only for a few hours are called short period marketing. The products dealt within these markets are of a highly perishable nature, such as fish, fresh vegetables and liquid milk. In these markets the prices of commodities are governed mainly by the extent of demand for rather than by the supply of the commodity.

(b)   *Long period markets* : These markets are held for a longer period than short-period markets. The commodities traded in these markets are less perishable and can be stored for some time, these are foodgrains and oilseeds. The prices are governed both by the supply and demand forces.

(c) *Secular markets* : These markets are of permanent nature. The commodities traded in these markets are durable in nature and can be stored for many years. Examples are markets for machinery and manufactured goods.

4. **On the basis of volume of transaction :** There are two types of markets on the basis of volume of transactions. Wholesale markets and Retail markets

   (a) *Wholesale markets* : A wholesale market is one in which commodities are brought and sold in large lots or in bulk. Transaction in these markets take place mainly between traders (etail sellers) on large quantity.

   (b) *Retail markets* : A retail market is one in which commodities are brought by and sold to the consumers as per their requirements. Transactions in these markets takes place between retailers and consumers. The retailer purchase in wholesale market and sell in small lots to the consumers. These markets are very near to consumers.

5. **On the basis of nature of transactions :** The markets which are based on the type of transactions in which people are engaged are of two types Spot or cash markets and Forward markets

   (a) *Spot or cash markets* : A market in which goods are exchanged for money immediately after the sale is called the spot or cash market.

   (b) *Forward markets* : A markets in which the purchase and sale of a commodity takes place at time 't' but the exchange of commodity takes place on some specified date in future, i.e., time t +1 sometimes even on the specified date in the future (t+1), there may not be any exchange of the commodity. Instead, the differences in the purchase and sale prices are paid or taken.

6. On the basis of number of commodities in which transaction taken place. A market may be general or specialised.

   (a) *General market* : A market in which all types of commodities such as foodgrains, oil seeds, fiber crops, gur etc. are brought and sold is known as general markets. These markets deal in a large number of commodities.

   (b) *Specialized markets* : A market in which the transactions takes place in one or two commodities is known as specialized market, for every group of commodities separate market exist. The examples are foodgrain markets, vegetable markets, wool market and cotton market.

7. **On the basis of degree of competition :** Each market can be placed on a continuous scale, starting from a perfectly competitive point to a pure

monopoly situation. Extreme forms are almost non-existent. Nevertheless it is useful to know their characteristics. In addition to these two extremes, various midpoints of this continuum have been identified. The markets are :

(a) **Perfect market**

(b) *Imperfect market* : 1. Monopoly market 2. Duopoly market 3. Oligopoly market, 4. Monopolistic competition

(a) *Perfect market* : Is one in which the following conditions hold good

1. There is a large number of buyers and sellers.,

2. All the buyers and sellers in the market have perfect knowledge of demand, supply and prices.

3. Prices at any one time are uniform as per a geographical area, plus or minus the cost of getting supplies from surplus to deficit areas.

4. The prices are uniform at any one place over periods of time plus or minus the cost of storage from one period to another.

5. The prices of different forms of products are uniform, plus or minus the cost of converting the product from one form to another.

Imperfect markets : The markets in which the conditions of perfect competition are lacking are characterized as imperfect markets.

Monopoly market : Monopoly market is a situation in which there is only one seller of a commodity. He exercises sole control over the quantity or price of the commodity. In this market, the price of a commodity is generally higher than in other markets. Indian farmers operate in a monopoly market. When purchasing electricity for irrigation. Where there is only one buyer of a product the market termed as a monopsony market - Cotton procurement in M.S.

Duopoly market : A duopoly market is one which has only two sellers of a commodity. They may mutually agree to charge a common price which is higher than the hypothetical price in a common market. The market situation in which there are only two buyers of a commodity is known as the duopsony market.

Oligopoly market : A market in which there are more than two but still a few sellers of a commodity is termed as an oligopoly market. A market having a few (more than two) buyers is known as oligopsony market.

Monopolistic competition : When a large number of sellers deal in heterogeneous and differentiated form of commodity, the situation is called monopolistic competition. The difference is made conspicuous by different trade marks on the product. Examples of monopolistic competition faced by farmers may drawn from the input markets. For example, they have to choose between various mark of insecticides, pump sets, fertilizers and equipment's.

8. **On the basis of nature of commodities :** Based on goods dealt in, market may be classified into Commodity and Capital market.

   (a) *Commodity market* : A market which dealt in goods and raw materials, such as wheat, barley, cotton, fertilizer, seed etc. are termed as commodity markets.

   (b) *Capital market* : The market in which bonds, shares, and securities are bought and sold are called capital markets for example money markets and share markets.

9. **On the basis of stage of marketing :** Markets may be classified into two categories. (a) Producing markets and (b) Consuming markets.

   (a) *Producing markets* : Those markets which mainly assemble the commodity for further distribution to other markets are termed as producing markets. Such markets are located in producing areas.

   (b) *Consuming markets* : Markets which collect the produce for final disposal to the consuming population are called consumers markets. Such markets are generally located in areas where production is inadequate ex. thickly populated urban centre.

10. **On the basis of extent of public intervention :** Based on the extent of public intervention, markets may be placed in any of following two classes. (a) Regulated markets (b) Unregulated markets.

    (a) *Regulated markets* : Markets in which business is done in accordance with the rules and regulations framed by the statutory market organization and represent different sections involved in the markets. The marketing costs in such markets are standardized and practices are regulated.

    (b) *Unregulated markets* : These are the markets in which business is conducted without any set rules and regulations. Traders frame the rules for the conduct of the business and run the market. These markets suffer from many ills, ranging from unstandardized charges for marketing functions to imperfections in the determination of prices.

SIZE OF THE MARKET - DEPENDS ON

1. **Character of commodity :** It is very wide covering the whole country or whole world when the commodity is portable, durable, suitable for sampling grading and supply can be increased. Example : The commodities of general consumption. A narrow market exists for perishable articles - fish, fruits

2. **Nature of demand :** Universal demand lead to wide market Restricted demand means narrow market.

3. **Communication and Transport :** developed system enable commodities carried long distance and establish wide contacts.

4. **Peace and security :** In war time markets get restricted.

5. **Currency and credit :** Developed policy lead to conveniently and profitable marketing on extensive area.

6. **Policy of the state :** Prohibitive duties and quotas restrict market zoning system.

7. **Degree of division of labor :** The greater the division of labour the cheaper the articles and wider the market.

PRICE DETERMINATION

The interaction/equilibrium between forces of demand and supply determine prices in the market. It is not demand of single buyer or supply of single seller but for all the buyers demand and sellers supply together that determine the price by their interaction. Therefore, the price of a commodity given for each individual consumer of firm is determined by all the consumers and firms that buy and sell.

Two Approaches to pricing under perfect competition

1. **Partial equilibrium approach set by Alfred Marshall :** Partial equilibrium analysis is based on the assumption of all other things in the economy are unaffected by any changes in the sector under consideration termed 'ceteris paribus'. This assumption is always violated to some extent.

2. **General Equilibrium Approach :** The general equilibrium analysis explains the mutual and simultaneous determination of all goods and factors. It thus looks at multi-market equilibrium, the prices of all goods are set simultaneously in its own flux.

The assumption violated and the effect is significant as in the case of inter-related goods, the partial equilibrium analysis ceases to be applicable. In such cases, we have to resort to the general equilibrium approach.

PRICE DETERMINATION : GENERAL STATEMENT

Marshall gave equal importance to both demand and supply in determination of the value or price. Marshall's famous analogy of pair of scissors is wroth quoting whether the upper or under blade of pair of scissors that cuts a paper. Neither the upper blade nor the lower one take separately can do the work separately, both have their importance in the process of cutting. The demand of all consumers and the supply of all firms together determine the prices which are then taken and given by each one of them.

The price is determined by the interaction of demand and supply. But the demand and supply are themselves governed by host of other factors. Professor Samuelson remarks, supply and demand are not ultimate explanation of price. They are simply useful catch-all categories for analysing and describing the forces causes and factor impinging on price.

Cost of production is main determinant of supply curve and so price Income of consumers and size of population determine demand and so the price. The firms seek to minimise cost and maximise profits. This govern resource use supplied by household sector- their willingness and ability to provide resources. Thus, price depends on the willingness and ability of the households to sell resources and buy goods. As well as the firms willingness and ability to sell goods and buy resources.

Marshall laid emphasis on role of time element in determination of prices. Because, supply condition vary with the length of period under consideration. On the basis of response of supply over time to a given and permanent change in demand three periods in which equilibrium between demand and supply was brought.

(i) *Very short period or market period equilibrium* : Supply is limited or fixed to the existing stock in hand

(ii) *Short run equilibrium* : When firms can expand output with the existing plants by changing the amounts of variable factors employed.

(iii) *Long run equilibrium* : When firms can abandon old plant or build new ones and new firms can enter the industry or old can leave it.

On the basis of time allowed to the forces of demand and supply to adjust the market the prices are market price, short-run price or long-run normal price. In very short run it is demand which determine price and in long run it is supply which determines prices. Thus, the factors which determine prices both demand and supply are important only their influence varies over different time periods.

Of Reproducible goods (nonperishable) The goods can be preserved or kept back from market and carried over to next market period. The supply curve can not be straight line. If the price is very high the seller will prepared to sell the whole stock and at very low price level, seller would not sell any amount and wait for better price in another market/market period, i.e., Reserve price : The price below which a seller will refuse to sell is called the reserve price (RP). The reserve price will depend on several factors.

Expectations regarding the future price - Higher RP in future

Liquidity preference - more urgent need for cash lower RP

Charges to for carrying stock-longer period/high charges lower will RP

Future cost - If the costs are expected to fall RP will be lower.

Durability of goods - the greater the durability higher is the RP.

Too much importance attached to cost incurred in past and fix higher RP. Given two extreme price levels.

1. The seller prepared to sell whole stock.

2. He will refuse to sell any the amount which he will offer for sale will vary with price. The supply curve of seller will slope upwards to right.

In short run, firm is in equilibrium at price equals marginal cost because fixed costs are disregarded in making decision to produce. It is average variable cost that determine to produce or not. If the price falls below AVC firm will shut down to minimize losses. Short run supply curve of industry is the lateral summation of short run marginal cost curve (MCC) of the firms. The supply curve of the industry lies above minimum AVC and it slopes upward from left to right since MCC slopes upward.

I. If there is an increase in demand the market price will rise sharply at level new demand curve intersect supply curve, supply of output remaining unchanged. Under the stimulus of increased demand firm will increase production. Hence in short run a larger amount of quantity is sold and the price is not quite as high as in the market period.

II. If there is decrease in demand, price will fall sharply the supply of the output remaining the same. But in short run firms will contract output by diminishing the variable factor and quantity supplied will decrease.

Thus short run price will be higher than new market price.

Price determination - Imperfect Competition - Monopoly

It is the market form in which single producer controls the whole supply of a single commodity which has no close substitute. The distinction between the firm and industry disappear under monopoly. The absence of close substitute means no other firms producing similar product/products ex electricity and water supply by local public utilities. This can also expressed as cross elasticity of demand - change in demand for a commodity as a result of change in price of another commodity. In monopoly the cross elasticity of demand between the product of monopolist and the product of any other producer must be very low.

Due to these condition the monopolist can set the price of his product and pursue an independent price policy. Thus essence of monopoly is power to influence price. However, only one at a time can be dictated either price or quantity demanded leaving the other to market forces.

BASES OF MONOPOLY

Barriers to the entry of rivals - economic in nature

Economies of large scale production ex. Automobile, aluminum and steel

Natural monopolies-in some industries competition is impracticable.

Artificial monopolies : These industries are given exclusive rights by govt., Exclusive ownership of raw material, Patent laws protect inventor. The entry of new competition blocked or eliminated by cut throat prices.

Price discrimination : Charging different price for the same product or the same price for differentiated product. Price discrimination may (a) personal, (b) local, (c) according to trade. Example : use electric supply for industrial, household & trader or firms.

Degree of price discrimination : In 1st degree different price for each unit of commodity, they charges the maximum that each buyer is able and willing to pay. This involves maximum exploitation of the buyers and known as perfect price discrimination.

In second degree the buyer are divided into groups and from each group a different price is charged which is the lowest demand price for the group.

In third degree discrimination the monopolist splits the entire market into few markets and charges a different price in each sub-market.

Conditions of price discrimination

1. When consumers have certain preferences on prejudices - elite/upper class people

2. When the nature of good - direct service goods example, beauty parlor

3. When consumer separated by distance or tariff barriers.

4. Govt. regulations - Electricity rates fixed for industrial user lower than domestic/commercial.

5. Ignorance and lethargy - ignorance of customers and disinclination to take trouble

6. Same service for different purposes - railway freight for coal, fruit machinery etc.

7. Special orders - for goods supplied to special order discriminating prices are charged.

Price discrimination is profitable? Only if elasticity of demand between the markets differ. The monopolist will charge more in the market with low elasticity and lower price where elasticity is higher. If demand curves are iso-elastic, then discrimination will not profitable.

Effect of price discrimination

From the analysis of price output equilibrium under discriminating monopoly we find (i) It increases the monopoly power of the producer. (ii) It gives the monopolist higher profits. (iii) The total output is larger than under simple monopoly.

Is price discrimination beneficial to society? No

It is desirable that price discrimination should be permitted in public supplies - so the poor can afford and the total receipts may be adequate to meet the total cost with profit. Thus every one may gain from production of such facilities when discriminatory prices are charged. Such discrimination involve raising the price for some people and lowering for some others. Thus the net effect on social welfare will be useful.

MARKETING FUNCTION AGENCIES AND INSTITUTIONS

Marketing function is activity performed in carrying product or movement of goods from producer to consumer. The marketing functions are as below:

Packaging : It means placing the goods in small packages. Packing means wrapping and crating of goods.

Transportation : Movement of products between places. It is an indispensable marketing function and its importance increase with urbanization.

Grading and standardization : Grading means sorting of the unlike lots according to quality specification. Standards are established on the basis of characteristics like size, shape, colour etc.

Labeling : The graded product indicate purity and quality of product based on standards fixed by Agricultural Marketing Adviser bear label 'AGMARK'

Storage and warehousing : Holding and preserving goods means storage. It adds time utility.

Warehousing : Is scientific storage structure to protect quality product.

Processing : Change in form of commodity. It include all manufacturing activities.

Buying and selling : It includes planning, contractual function, negotiation of price, agreement and transfer of goods.

Market information : Communication or reception of knowledge.

Financing : Credit is the lubricant that facilities the marketing machine.

Future trading : Buying and selling in future market of equal and opposite quantity helps to avoid risk and uncertainty.

Marketing Agencies and Institutions

Agro-climatic variation limit production of farm products to certain area which needs movement from producer to consumer by direct route or an indirect route. In case of direct route complete absence of middlemen and very small proportion of farm produce handled. Indirect route involves intermediaries or middlemen. In modern era of specialized production both horizontal and vertical distance between producer and consumer has increased. The agencies are (a) Merchant middlemen (i) Wholesaler (ii) Retailers (b) Agent middlemen (c) Speculative middlemen (d) facilitative middlemen

1. **Producer :** sell surplus in village or in market. Some farmers assemble the produce of small farmers and make profit. Constant touch with market bring home to them a fair knowledge of market practices. An access to market information able to perform the functions of market middlemen.

2. **Middlemen :** Specialized in market functions and render services. They do this at different stages in the marketing process.

   (a) *Merchant middlemen* - Take title to the goods they handle. Buy and sell on their own and gain or loss depending on prices. Wholesalers buy and sell in large quantities from farmer or trader. They do not sell significant quantities to ultimate consumers.

      (i) *Wholesalers :* Assemble goods in lots of different quality from farms and prepare for market. Equalize flow of goods by sorting them in peak arrivals and releasing them in off season. Regulate flow of goods by trading in various markets. Finance the farmer to meet the requirement of production inputs. Assess demand of prospective buyer and processors and plan movement of goods over space and time

      (ii) *Retailers :* Buy from wholesaler and sell to consumer in small quantities. They are producers personal representations to consumers. Itinerant trader and village merchants are retailers. Itinerant traders are petty merchants who move from village to village and directly purchase the produce from the cultivators.

   (b) *Village merchants :* Supply finance and essential goods to farmers and purchase farm produce to sell in market or retain for later sale.

3. **Agent middlemen :** Act as representative of their clients. They do not take title and therefore do not own it. They merely negotiate the purchase and or sale. They sell services to their principals and not goods. They receive income in the form of commission or brokerage. Agent middlemen are of two types (1). Commission agent or Arhatias and (2). Brokers. Commission Agent - operate in wholesale market act as representative of buyer or seller. He is usually granted broad powers by those who consign goods or order the purchase. He normally takes over

physical handling of produce arrange for its sale deduct his expenses and commission and remit the balance to the seller.

Commission agent in unregulated market are **Kaccha arhatias** and **Pacca arhatias.**

Kaccha arhatias primarily act for the sellers including farmers. They provide advance to farmers and itinerant traders on condition to sell through them and charge commission in addition to normal rate of interest for advance.

Pacca arhatia act on behalf of the traders in consuming market. The processor and big wholesaler in consumer market employ pacca arhatia as their agent for purchase of specified good within given price range.

In regulated market : 'A' class trader is commission agent keep an establishment like shop, godown and rest house for his clients. He render all facilities. Therefore preferred by farmers than the Cooperative Marketing Society

FUNCTIONS OF COMMISSION AGENT

Advance 40-50% of expected value of crop as loan.

They act as bankers of the farmer, retain sale proceeds and pay as require.

Offer advice to farmer for purchase of input and sale of products.

Provide empty bag to bring produce to market.

Provide facilities like storage, transport, accommodation and food to farmer as well help in personal difficulties.

Broker : Render personal service to their clients in the market.

They bring together buyer and seller for negotiation and do not have physical control of product on establishment. Their charge is called brokerage. The number of broker in foodgrain marketing is declining but still play a valuable role in gur, sugar, oil, cottonseed and chillies. In most regulated markets broker do not pay any role because of open auction.

Speculative middlemen : Take title to the product with view to making a profit on it. They specialized in risk taking. They make profit from price fluctuations.

Facilitative middlemen : Assist in marketing, Increase efficiency by their work, and receive income as fee from service user. Hamal/Labour - Physically move products, unloading, loading, weighing cleaning, sieving, refilling bags and stitch the bags.

Hamal are hub of the marketing wheel.

Weighmen : Pan balance for small quantities. General scale beam balance used. The weighbridge system of weighing exist only in few big markets.

Marketing Institutions : Are business organization which come up to operate the marketing, e.g., Individual, cooperative and Govt. institutions

Important institutions in agril. marketing are

State Trading Corporation (STC),

Food corporation of India (FCI),

National Agril. Cooperative Marketing Federation (NAFED),

Cotton Corporation of India (CCI),

Jute corporation of India (ICI),

National Dairy Development Board (NDDB),

National oilseed and vegetable oil Development Board (NOVODB), Tobacco Board,

Agricultural Processed Products and Export Dev. Agency (APEDA)

The Directorate of Marketing and Inspection, Govt. of India

State Level Agril. Marketing Department,

Maharashtra State Agril. Marketing Board (MSAMB)

State and lower level cooperative marketing societies, fair price shops, consumers cooperative stores, milk unions.

Marketing channel : Are routes through which agricultural products move from producers to consumers. It involves above intermediaries they put their efforts and incur costs to carry out the different marketing functions. It extends from producer to consumer. Under the traditional system of marketing of the agricultural products, producer-sellers incurred a high marketing cost, and suffered from unauthorized deductions of marketing charges and the prevalence of various malpractices.

REGULATION OF AGRICULTURAL MARKETING

To improve marketing conditions, with a view to create fair competitive conditions to increase the bargaining power of producer-sellers was considered to be the most important pre-requisite of orderly marketing. Most of the defects and malpractices under the marketing system of agricultural products have been more or less removed by the exercise of public control over markets, i.e., by the establishment of regulated markets in the country.

Definition : A regulated market is one which aims at the elimination of the unhealthy and unscrupulous practices, reducing marketing charges and providing facilities to producer-sellers in the market. Any legislative measure designed to regulate the marketing of agricultural produce in order to establish, improve and enforce standard marketing practices and reasonable charges may be termed as one which aims at the establishment of regulated markets.

Regulated markets have been established by State Governments, and rules and regulations have been framed for the conduct of their business. The establishment of regulated markets is not intended at creating an alternative marketing system. The basic objective has been to create conditions for efficient performance of the private trade, through facilitating free and informal competition. In regulated markets, the farmer is able to sell his produce in the presence of several buyers through open and competitive bidding. The legislation for the establishment of regulated markets does not make it compulsory for the farmer to sell his produce in the regulated market yard. Instead, voluntary action on the part of the farmers to take advantage of such a market is assumed. The basic philosophy of the establishment of regulated markets is elimination of malpractices in the system and assignment of dominating power to the farmers or their representatives in the function of the markets.

Objectives : The specific objectives of regulated markets are :

(i)   To prevent the exploitation of farmers by overcoming the handicaps in the marketing of their products.

(ii) To make the marketing system most effective and efficient so that farmers may get better prices for their produce and the goods are made available to consumers at reasonable prices;

(iii) To provide incentive prices to farmers for a better production programme, both in quantitative and qualitative terms; and

(iv) To promote an orderly marketing of agricultural produce by improving the infrastructural facilities.

IMPORTANT FEATURES OF REGULATED MARKETS

Under the provisions of the Agricultural Produce Market Act, the state government gives notice of its intention to bring a particular area under regulation by notifying the market are, market yard, main assembling market and sub-market yard, if any, under the principal regulated market. The meaning of these terms is explained in the following paragraphs.

(i)   **Market Area :** The area from which the produce naturally and abundantly flows to a commercial center, i.e. the market, and which assures adequate business and income to the market committee.

(ii) **Principal Assembling Market :** It is the main market which is declared as a principal market yard on the basis of transactions and income generated for the market committee.

(iii) **Sub Market Yard :** It is the sub-yard of the principal assembling market. This is a small market and does not generate sufficient income to be declared as a principal assembling market.

(iv) **Market Yard :** This is a specified portion of the market area where the sale, purchase, storage and processing of any of the specified agricultural commodities are carried out.

Some of the important features of regulated markets are :

Methods of sale : In regulated markets, the sale of agricultural produce is undertaken either by open auction or by the close tender method. These sales methods ensure a fair and competitive price for the produce and prevent the cheating of farmers by market functionaries. By these methods, the sale is carried out under the supervision of an official of the market committee and the signature of the buyer are taken as soon as the auction is over. The business is done during fixed hours. Generally, different commodities are traded at different times of the day. Bidding is kept open to all authorized traders and the highest bider is given the produce. The farmer has the option to refuse or to accept the sale even at the highest bid, if he so desires.

Weighment of Produce : Weighment of the produce is done by a licensed weighman with standard weights and a platform scale. In some markets, a weighbridge has been installed. This eliminates short weights and malpractices which arise out of weighing with a hand balance.

Grading of Produce : The produce in the regulated markets is sold only after grading; but because of the absence of facilities, such as space, funds for the employment of technical experts for grading and the purchase of grading equipment, the grading scheme could not be implemented in all the markets in the country. The scheme is in operation only in some markets. According to the National commission on Agriculture only 13 per cent of the regulated markets have grading facilities. There is, therefore, a need for the extension of grading facilities to all the markets in the country.

Market News Service : In regulated markets, there is an arrangement for a proper and correct dissemination of market prices, through various media, such as loudspeakers and notice boards. However, the dissemination of price information is not perfect in most markets because of the non-existence of these facilities. Sometimes, farmers are unable to take advantage of this facility either because of their illiteracy or the non-availability of information when they require it.

Market Charges : With the regulation of markets, such market charges as darmada, charity, karda, dhalta and muddat, were abolished, while the rates of other market charges, such as commission, brokerage, hamali and weighing charges, were specified in proportion to the extent of the service rendered by middlemen. Recently, the market charges payable by sellers have been transferred to the buyer; the seller have to pay only for the activities undertaken prior to marketing of the produce, i.e., for transportation and hamali only. There is, thus, some saving for the producers. The extent of saving varies from market to market and from commodity to commodity.

Payment of the value : It is obligatory on the buyer to make prompt payments for the produce without deductions of any muddat. In unregulated markets, the payment of the price of the produce is made by the buyers after 3 to 15 days of the sale of the produce in accordance with local market rules. In the past, if the seller insisted on cash payment, the buyer would deduct muddat (cash discount) to the extent of Rs. 0.50 for the produce worth Rs. 100 while making the payment. This practice has been stopped with the regulation of markets.

Licensing of Market Functionaries : All the market functionaries, from the hamals to traders working in the regulated market, have to obtain a license from the market committee, after paying the prescribed fees, to carry on their business. The licensed traders have to keep proper records and maintain accounts in accordance with the bye laws of the market committee. This facilitates the exercise of a proper control on the accounts of the traders and the collection of the correct amount of the market fee by the market committee. Any violation of rules by middlemen may lead to the cancellation/suspension of the license by the market committee or to the filing of challans in a court of law.

Supervision : The day-to-day functioning of regulated markets is supervised by the officials of the market committee, i.e., the secretary, auction clerks, and other staff. The administrative decisions are taken by the nominated/elected market committee.

The market committee consists of representatives of all sections, i.e., farmers, traders, co-operative marketing societies, co-operative or commercial banks, autonomous bodies (Panchayat Samiti and Muncipal Board of the area) and government officials. Prior to the establishment of regulated markets, the rules for the conduct of the business in the market were framed by traders without any consideration of interests of other groups of persons, i.e., farmers and consumers.

The market committee is nominated in the first instance for a period of two years. Subsequent committees are elected for a term of three years each. The members of the market committee elect their own Chairman and Deputy Chairman. The market committees are the main instruments under the Market Act.

CO-OPEATIVE MARKETING

The establishment of cooperative marketing societies is another step which has been taken to overcome the problems arising out of the present system of marketing agricultural produce. The objectives of economic development and social justice can be furthered by channelising agricultural produce through co-operative institutions.

Meaning : A co-operative sales association is a voluntary business organization established by its member patrons to market farm products collectively for their direct benefit. It is governed by democratic principles and savings are apportioned to the members on the basis of their patronage. The members are the owners, operators and contributors of the commodities and are the direct beneficiaries of the savings that accrue to the society. No intermediary stands to profit or lose at the expense of the other members.

Definition : Co-operative marketing organizations are associations of producers for the collective marketing of their produce and for securing for the members the advantages that result from large-scale business which an individual cultivator cannot secure because of his small marketable surplus. In a co-operative marketing society, the control of the organization is in the hands of the farmers, and each member has one vote irrespective of the number of shares purchased by him. The profit earned by the society is distributed among the members on the basis of the quantity of the produce marketed by him. In other words, co-operative marketing societies are established for the purpose of collectively marketing the products of the member farmers. It emphasizes the concept of commercialization. Its economic motives and character distinguish it from other associations. These societies reassemble private business organization in the method of their operations; but they differ from the capitalistic system chiefly in their motives and organizations.

Functions of co-operative marketing societies are :

(*i*)   To market the produce of the members of the society at fair prices;

(*ii*)   To safeguard the members from excessive marketing costs and malpractices;

(*iii*) To make credit facilities available to the members against the security of the produce brought for sale;

(*iv*)   To make arrangements for the scientific storage of the members' produce;

(*v*)   To provide the facilities of grading and market information which may help them to get a good price for their produce;

(*vi*)   To introduce the system of pooling so as to acquire a better bargaining power than the individual members having a small quantity of produce for marketing;

(*vii*) To act as an agent of the government for the procurement of foodgrains and for the implementation of the price support policy;

(*viii*) To arrange for the export of the produce of the members so that they may get better returns;

(*ix*)   To make arrangements for the transport of the produce of the members from the villages to the market on collective basis and bring about a reduction in the cost of transportation; and

(*x*)   To arrange for the supply of the inputs required by the farmers, such as improved seeds, fertilizers, insecticides and pesticides.

TYPES OF CO-OPERATIVE MARKETING SOCIETIES

On the basis of the commodities dealt in by them, the co-operative marketing societies may be grouped into the following types:

(i) **Single commodity : Co-operative Marketing Societies :** They deal in the marketing of only one agricultural commodity. They get sufficient business from the farmers producing that single commodity. The examples are Co-operative Sugar, Cotton Co-operative Fruit Sale Cooperative Marketing Society etc.

(ii) **Multicommodity : Co-operative Marketing Societies :** They market a large number of commodities produced by the members, such as foodgrains, oilseeds, vegetable, fruits etc..

(iii) **Multi-purpose, Multi-commodity : Co-operative Marketing Societies :** These societies market a large number of commodities and perform such other functions as providing credit to members, arranging for the supply of the inputs required by them and meeting their requirements of essential domestic consumption goods.

STRUCTURE

The co-operative marketing societies have both two tier and three tier structure. In the states of Assam, Bihar, Kerala, Madhya Pradesh, Karnataka, Orissa, Rajasthan and West Bengal, there is a two tier pattern with primary marketing societies at the taluka level and state marketing federation as apex body at the state level. In other states, there is three tier system with district marketing society in the middle. At the national level, NAFED serves as the apex institution. The pattern of the three tier structure is as below:

(i) Base Level : At the base level, there are primary co-operative marketing societies. These societies market the produce of the farmer members in that area. They may be single commodity or multi-commodity societies, depending upon the production of the crops in the area. They are located in the primary wholesale market and their field of operations extends to the area from which the produce comes for sale, which may cover one or two tehsils, panchayat samities or development blocks.

(ii) Central/District Level : At the district level, there are central co-operative marketing unions or federations. Their main job is to market the produce brought for sale by the primary co-operative marketing societies of the area. These are located in the secondary wholesale markets and generally offer a better price for the produce. The primary co-operative marketing societies are members of these unions in addition to the individual farmer members. In the two tier structure, the State societies perform the functions of district level societies by opening branches throughout the district.

(iii) State Level : At the State level there are apex (State) co-operative marketing societies. These State level institutions serve the State as a whole. Their members are both the primary co-operative marketing societies and the central cooperative unions of the State. The basic function of these is to co-ordinate the activities of the affiliated societies and conduct such activities as inter State trade, export import, procurement, distribution of inputs and essential consumer goods, dissemination of market information and rendering expert advice on the marketing of agricultural produce.

The cooperative marketing network of the country includes 29 state level marketing federations, 173 district/regional marketing cooperative societies, 2633 general purpose primary marketing societies and 3290 special commodity societies.

Membership : There are two types of members of co-operative marketing societies :

(i) *Ordinary Members :* Individual farmers, co-operative farming societies and service societies of the area may become the ordinary members of the co-operative marketing society. They have the right to participate in the deliberations of the society, share in the profits and participate in the decision making process.

(ii) *Nominal Members :* Traders with whom the society establishes business dealings are enrolled as nominal members. Nominal members do not have the right to participate in decision making and share in the profits.

Functions : Carried out by the co-operative marketing societies are :

(i) **Sale on Commission Basis :** Co-operative marketing societies act as commission agents in the market, i.e., they arrange for the sale of the produce brought by the members to the market. The produce is sold by the open auction system to one who bids the highest price. The main advantage, which the farmer-members get by selling the produce through co-operative marketing societies instead of a commission agent is that they do not have to accept unauthorized deductions or put up with the many malpractices, which are indulged in by individual commission agents. As there is no individual gain to any member in the marketing of the agricultural produce through co-operative marketing societies, no malpractice's are expected to be indulged in.

This marketing is not risky for co-operative societies. But sometimes traders in the market form a ring and either boycott the auction or bid a low price when the produce is auctioned on the co-operative marketing societies shops. These tactics of the traders reduce the business of cooperative marketing societies. Therefore, farmers hesitate to take their produce for sale in the market through co-operative marketing societies.

(ii) **Purchase of Members' Produce :** Co-operative marketing societies also enter the market as buyers. A society participates in bidding together with other traders, and creates conditions of competition. The commodities thus purchased by a society are sold again when the prices are higher. This system of the outright purchase of the produce by the society involves the risk of price fluctuations. If the managers of societies lack business experience, they hesitate to adopt the outright purchase system. In 1964-65, the National Co-operative Development Corporation recommended that the outright purchase system should be adopted only by a society which possesses the following qualities:

   (a) The society has a trained manager, i.e., one who is capable of understanding the intricacies of the trade;

   (b) The society is financially sound and has adequate borrowing facilities;

   (c) The society is affiliated to a good central level society; and

   (d) The society possesses processing facilities.

(iii) **Advancement of Credit :** Co-operative marketing societies advance finance to farmers against their stock of foodgrains in the godowns of the societies. This increases the holding power of the farmers and prevents distress sales. Generally, societies advance credit to the extent of 60 to 75 per cent of the value of the produce stored with them. The recoveries are effected from the sale proceeds of the produce of the farmer. This function involves no risk to the society, it increases the business.

(iv) **Procurement of Foodgrains :** Co-operative marketing societies act as agents of the government in the procurement of foodgrains at the procurement prices.

(v) **Other Functions :** The following functions are also carried out by them, depending upon the availability of funds and other facilities. They assemble the marketable surplus of small and marginal farmers and transport this surplus from villages to the society headquarter for disposal. They make arrangements for the grading of the produce and encourage producers to sell the produce after grading so that they may get better prices. They undertake the processing of produce. They make arrangements for the export of agricultural commodities.

FARM MANAGEMENT AND PRINCIPLES

Meaning : Farm is unit of land. Farm is one or more tracts of land held or operated as a unit for the production of agricultural products. Farm means piece of land where crop and livestock enterprises taken up under common management and has specific boundaries. Management- act or art of managing, Farm - Socio-economic unit provide income and happiness to farm families. It is decision making unit where resources have alternative uses need to increase output by adding input - combine crop and livestock. With limited resources raising agricultural productivity depend largely on better management. Each farm consist of tangible land, labour, capital and intangible factor management. Management guide use of tangible factors.

Thus successful operation of farm depends on sound management technique. It is technique in making decision for choosing alternative use of available limited farm resources and for operating farm plan to maximize agricultural production and also economic returns for the farm as a whole. It is rational resource allocation of scarce farm resources towards set of goal for maximum family satisfaction.

Farm management : Defined as 1 'Science which deals with judicious decision on the use of scare farm resources having alternative uses to obtain maximum profit and family satisfaction continuous from farm as a whole and under sound farming programme.

2. According to Gray - The art of managing a farm successfully as measured by the test of profitableness is called farm management.

3. Farm management is the study of business principles in farming-Warren.

4. Efferson defined - The science which considers organization & operation of farm from point of view of efficiency & continuous profit.

5. Hadelson defined - Management in farming/business consist chiefly in making correct decision at the right time & then seeking that decision are carried to successful completion.

Farm Management : Is science of decision making, managing farm is continuous process of decision making. The need for it arises from changes occurring in the farm conditions and outside farm. Correct decision and quick adjustability need for profitability and survival. Farm management is branch of agricultural economics which deals with farming principles and practices with an objective of obtaining maximum possible returns from the farm. Farm is unit of land on which farmer does, the planning, furnishes supervision, own power and equipment and provide labour.

OBJECTIVES OF FARM MANAGEMENT

To obtain maximum and continuous net profit from the enterprises on farm. Farm management gives knowledge to analyze business.

To study input output relation and efficiency of factor combination.

It helps to evaluate farm resources and land use.

To study cost of cultivation, cost of production and comparative economics of enterprises.

To study impact of technical changes on farm business and find ways/means for increasing efficiency of farm business.

SCOPE OF FARM MANAGEMENT

Farm management is considered to fall in the field of micro-economics. It deals with allocation of resources at individual farm level. Aspects of farm business influencing economic efficiency of farm are covered by the farm management studies.

The subject of farm management include research, training, teaching and extension. Farm management research provide solution to economic problems faced by farmers. Training and teaching help to take decision about what to produces? how to produce? and when to produce? Farm management extension improves managerial ability of the farmers.

FARM MANAGEMENT PRINCIPLES

Farm management principles serve as a guideline for collecting and using requisite information for rational decision making of farm business. They also provide a set of tools for the preparation of farm budgets and production programmes. These principles help provide answers to various farm problems and save time and energy otherwise lost in trial and errors to arrive at appropriate decisions. Knowledge of these principles enhances farm entrepreneurs' sense of judgment, a pre-requisite for meeting the demands of a business, especially under odd situations requiring a special intellect. The following are the six basic principles involved in making rational farm management decisions.

(i)   Principle of variable proportions or laws of returns.

(ii)   Cost principle.

(iii)   Principle of substitution between inputs (Least cost combination).

(iv)   Equi-marginal returns principle or opportunity cost principle.

(v)   Principle of substitution between products.

(vi)   Principle underlying decisions involving time and uncertainty.

These farm management principle are briefly discussed below :

1. Principle of variable proportions or laws of returns

It has three phases : (a) Diminishing returns, (b) Constant return and (c) Increasing returns.

(a) Diminishing returns : It is a basic natural law affecting many phases of management of a farm business. It is a law of fundamental importance in agriculture. This law describe the relationship between output and a variable input where other inputs are held constant. The law can be stated as follows : "If increasing amounts of one input are added to the production process while all other inputs are held constant, the amount of output added per unit of variable input will eventually start decreasing." It states that if the quantity of one factor increased with quantities of other factors held constant, the marginal increment to the total product may increase or remains constant at first but will eventually decrease after a certain point. The operation of this law can be, however, delayed by improvements in technology and/or improvement in managerial ability. Ultimately this law must operate in the practical world. The level to which yields per acre, milk per cow or weight per poultry bird should be pushed are the kind of questions which involve the law of diminishing returns. It is, thus, an important point in farming to decide the level to which a farmer should push his output per acre or per cow, etc. to secure the maximum possible profit. This principle of returns is also important in specifying how large a farm should be or how much labour and/or machinery be added. In this context resources can be classified as variable resources and fixed resources. For example, an acre of land would be a fixed resource with the farmer, but fertilizer would be a variable resource. As additional quantities of fertilizer are given to an acre of crop, the return to each additional dose will eventually become lesser and lesser. When diminishing returns hold true, it is seldom profitable to produce a maximum yield per acre or milk or meat per animal, although exceptions might exist.

It can be said that the added quantity of a variable resource applied to a fixed acre of land or given head(s) of livestock adds less and less to the yield or output. Examples are application of seeds, fertilizers, irrigation, etc. which have a characteristic of diminishing marginal productivity. There are some farmers who lose sight of diminishing returns to variable factor - use and consider the highest yield per acre, the highest milk yield per cow, etc. always to be the best level in terms of profit. This way they think only in terms of physical yields and not in terms of costs and profit. True, many farmers would need to raise their production levels in order to increase their profits, but they must consider cost at some point. One can easily decide the level of resource use or level of production by using the following profit rule under the conditions of diminishing returns. Keep adding variable resource(s) to the fixed resource(s) as long as the added return is more than the added cost. Such simple exercises for taking decisions of day to day operations on farm can save the farmer from many losses and increase his net returns from the farm business. This principle, should be therefore, helpful in making decisions such as :

1. The level to which yield per acre, milk per cow, etc. should be pushed to secure maximum profit.

2. The size of the farm one should operate with given resources of capital, labour and management.

3. The amount of fertilizer, labour or type of machinery one should use.

(b) Constant productivity : It means each marginal unit of variable resource adds the same amount of the output to the total production. Though 'diminishing marginal productivity' is the rule, constant productivity is frequently observed when no resource in fixed and all are increased together in the same proportion. For example, another acre may be as productive as the first with same inputs. If one acre of wheat requires 20 man-hours of labour, 30 Kgs. of seed and 13 inches of irrigation water and yields 10 quintals of wheat, the second acre will require an additional 20 man-hours of labour, 30 kgs. of seed and 13 inches of irrigation water and will also yield 10 quintal of wheat. The second acre is just as productive as first. The marginal or added production from each increase in resource input is the same; this is a case of constant productivity.

Another case is when one or more resources are fixed but have excess capacity. For example, family labour or a farmer may not be fully employed. A storage godown may have surplus capacity. A tractor may be big enough to control 50 acres holding but the farmer may have only 27 acres. If variable input is added to such a resource-mix situation, constant returns will result.

Under constant productivity, each unit input increase is just as profitable as another. Under such conditions the profit rule is: If production is profitable on first unit, keep producing till the constant returns hold. Do not produce at all, if production is not profitable on first unit. In a sense, follow the same principle, i.e., continue adding the variable resource to the fixed resource(s) as long as the return is greater than the added costs.

Limits on constant returns are reached as some of the factors become fixed. If nothing else becomes fixed, management becomes a fixed resource. The productivity of one resource depends on the amount of the other(s) with which it is used. For example, if capital is fixed at a low level for the farm as a whole, labour productivity will be lower. Since the productivity of one resource depends on the amount of other resource(s) with which it is combined, farmers having different quantities of land, capital, labour and management will have different programmes. What is best for one farm is therefore, seldom best for the other(s). Each farmer must get the right balance of resources and a unique optimum farm organization consistent with the resources he has.

(c) Increasing productivity : There are few cases in farming business where increasing productivity may be found. Increasing productivity means added resources give increasing returns. This relationship may hold only over a very limited range of production and is applicable when all resources are increased together and not when some resources are fixed. For example, a cattle shed constructed for 30 cows may cost more per cow than if one is constructed for 60 cows, The cost involved in the latter case may not be double because of some economies on account of joint walls etc., but the gross returns per cow might be the same. Use of added resource(s) thus, will give increasing returns in such cases. In this case each additional unit gives higher and higher returns. So long as this relationship holds, production should keep expanding.

2. Cost Principle : Most of the producers give considerable importance to the cost of production while producing a commodity, It generally refer to the expenses incurred in producing a unit of a product in a particular time period. Without specifying the amount and the time period, any reference to cost will be meaningless.

Accounting periods : There can be two accounting or planning periods: short-run and long-run. Short-run is a period of time which is sufficient enough to permit desired changes in output without altering the size of the plant. The long-run is generally considered to be the period which is sufficiently long for output to be altered by varying either the size of the plant or by making a more or less intensive utilization of the existing plant During the short-run, it may be possible for example to increase production through increased use of labour and/or by increasing the quantities of other inputs. In the long run, size of the holding can be varied either by increasing the acreage or by increasing the number of permanent workers. Thus, the manner in which production can be varied depends upon the length of the planning period under consideration.

Types of Costs : The relationship of costs to the income is very important in farming. As there are two categories of planning periods, there are two corresponding categories of costs viz., fixed costs and variable costs. Fixed costs are sunk costs or overhead charges. They do not change with the level of production. These costs are fixed such as, taxes, rent, electricity meter charges, insurance, depreciation, family labour, etc. These are the expenses on farm which must be paid even if nothing is produced. For example, if you hire a man on a year round basis, his wage is a fixed cost. Costs do not become fixed until they have been incurred, they do not vary with the change in output and can have no bearing upon management decisions with respect to using less or more of these inputs. In the short-run some costs are fixed and others can be varied. In the long-run, however, all costs become variable and influence long run decisions to stop, decrease or increase the size of the firm.

Variable costs are those which vary with the level of production. These include the costs of adding the variable inputs. They do not occur if we produce nothing: amount depends on what we produce. For example, if we use more fertilizer to produce more grains, the fertilizer costs go up. In making production decisions in the short-run to decrease or expand production, only variable costs need to be considered. Seed, tractor fuel, repairs, feed, fertilizer cost, other such items represent variable costs. Labour if hired on daily basis is also included in variable costs. If harvesting is done on custom basis, it is a variable cost.

Fixed costs plus variable costs are equal to total costs. Net revenue is equal to total revenue minus total costs. A farmer may have to quit farming if total revenue is not greater than total costs in the long-run. The costs relevant for a particular decision depend upon the nature of the decision under consideration.

Application of the Fixed and Variable Cost Principle

In order to maximize net revenue, variable (or added) costs are the relevant costs to be considered. In this context, profit rules for decisions in the long-run and short-run planning period can be described as under:

In the short run, gross return must cover the variable costs. The maximum net revenue is obtained when marginal cost (MC) equals the price of the product (MR).

If gross returns are less than total costs (variable + fixed costs) but are still larger than the variable costs, guiding principle should be to keep increasing production as long as added returns (MR) are greater than added costs (MC).

In the short run, MC-MR point may be at a level of input use which may involve a loss instead of profit. Yet, at this point loss will be minimized. In such a situation, objective should be to minimize losses in the short run. This situation of operating the farms when price or MR is greater than average variable cost but less than average total costs is common in agriculture. This explains why the farmers do as they do: why they keep farming, even when they run into losses.

In the long run, gross return should be more than variable plus fixed cost (total costs). For taking production decisions in such a situation one should go on using resources as long as the added returns remain greater than added total costs. Here the objective is to shoot at maximum profit instead of minimizing the losses.

The selling price must be, therefore, greater than the variable cost of producing each unit of product in the short-run farming is to be carried on. Over the long period however selling price must be greater than the variable cost and fixed cost per unit. Once these costs are covered, farm entrepreneur may not be interested in the average or per unit of cost of production. The marginal or added cost of each increase in production will be then important in determining how far he should go. It will be seldom economical to produce with the lowest per unit cost. The point of lowest average cost is not necessarily, the point of lowest marginal cost and even lowest marginal cost is not the optimum point. The optimum point will be where added returns are equal to the added cost. However, the optimum point can be at the point of lowest per unit cost only in a case where the minimum cost coincides with the maximum profit output. This is the case when selling price and marginal revenue are equal to minimum cost.

3. PRINCIPLE OF FACTOR-SUBSTITUTION
(Least-cost Combination)

In agriculture, various inputs or practices can be substituted in varying degrees for producing a given output. A producer has to choose a particular combination of inputs which would be most profitable.

At a given level of output, he has to decide upon the least cost-combination of practices or inputs. There are large number of alternatives for performing different farm operations and obtaining output through different combinations of inputs. A few of such alternatives can be (1) use of bullocks vs. tractor for a given size of a farm, (2) harvesting of crops by machines vs. by manual labour, (3) milking machines vs. hand milking for a given herd of cows, (4) wheat bhusa, green fodder and grains-mix in dairy feeds, (5) combinations of potash, phosphate and nitrogen in fertilizers for crops.

A number of combinations of concentrates and green fodder, for example, can be used in producing a given amount of milk, forages usually substitute at diminishing rate for grain. Problem there is to find the least cost combination of green fodder and grains, milk production remaining the same or otherwise, a combination of fodder and grains which, cost remaining the same, will yield maximum output of milk.

Cost minimization will not depend only upon the cost of inputs and prices of products but also on the rate of substitution. For example, if hand labour costs less relative to cost of performing the operations by machines, costs may be lowered by substituting labour for machinery. if machine costs are low relative to labour, labour should get substituted with machine operations.

Procedure for Working Out the Least-Cost Combination

Step 1 : Compute the substitution ratio (marginal rate of substitution) by dividing the number of units of the replaced resource by the number of units of the added resource:

$$\text{MRS} = \frac{\text{No. of units of replaced resource}}{\text{No. of units of added resource}} = \frac{\Delta X_1}{\Delta X_2}$$

Step 2 : Compute the price ratio by dividing the price of the "added" resource by the price of the "replaced" resource:

$$\text{PR} = \frac{\text{Cost per unit of added resource}}{\text{Cost per unit of replaced resource}} = \frac{Px_2}{Px_1}$$

Step 3 : Find the point where the substitution ratio and price ratio equals:

$$\frac{\Delta X_1}{\Delta X_2} = \frac{Px_2}{Px_1} \quad \text{Or} \quad \Delta X_1 . Px_1 = \Delta X_2 . Px_2$$

Profit Rules

(i) If the substitution ratio is greater than the price ratio, one can reduce the costs by using more of "added" resource.

(ii)   If the substitution ratio is less than the price ratio, costs can be reduced by using more of "replaced" resource.

(iii) If the substitution ratio equals the price ratio, it is the point of least cost.

Application of the Principle of Factor–Substitution

The rate at which the inputs can be exchanged to maintain a given level of output may be either constant or varying. Given the technical or physical rate of substitution, the least cost combination of resources will differ with different prices.

Special Cases of Substitution

There are certain practices which substitute for other and also add to the output at the same time. An example is hybrid maize seed replacing 'desi' maize seed. Hybrid maize uses about the same quantity of seed but produces more yield per acre. Here it may increase the cost but it adds more to value of the production than it adds to the cost. The superiority of the substitute practice is established not only by the cost of the new practice but also by the value of the added yield. In estimating the profitability of such practices, budgeting technique becomes more useful. It is a case of shift in the production function.

LAW OF EQUI-MARGINAL RETURNS

If the farmer had access to any amount of working capital and labour and could expand his acreage and building facilities etc., as far as he wished, he would not have any difficulty in deciding which commodities to produce: he would not compare one enterprise with another. Instead he would produce all crops and livestock products physically possible in his locality, considering weather, climate, soil and other physical factors. The decision rule would be simple: select all products which can be produced and expand output as long as the added returns are greater than added costs. But choice on the number and size of enterprises in this manner cannot be made because resources are limited. Expansion of one enterprise or practice generally requires an equivalent contraction in another. The question is which enterprise or combination of enterprises will give the greatest income? Such an optimum choice of enterprises is made based on the principle of equi-marginal returns or the principle of opportunity cost. Profits will be the greatest if each unit of labour, capital and land is used where it adds the most to the returns. In other words, this principle lays down: the best combination of enterprises or practices will be where limited resources are allocated in the manner that one cannot change the use of a single unit without reducing the income. This principle, thus, states that resources should be used where they bring not the greatest average returns, but the greatest marginal returns.

The principle dictates that the resources should be used not where they bring the highest average returns, but where they yield the highest marginal returns. The best combination of enterprises is attained not when we select profitable crops, but when we select the most profitable crops. The profitability of an enterprise depends on the price of the product, the direct costs attached to the enterprise and the amount of product sacrificed as one enterprise gets replaced with the other.

For the application of this principle, one needs data on prices of products, direct costs attached to each enterprise and also amount of production sacrificed when one enterprise is replaced with the other. Budgeting or programming techniques take this principle into consideration for working out an optimum production plan.

OPPORTUNITY COST PRINCIPLE

The farm resources are always limited and there are more than one alternative to use these resources. When resources are used in one product, some alternative is always foregone. The opportunity cost is the value of the next best alternative foregone. The value of one enterprise sacrificed is the cost of producing another enterprise.

The opportunity cost principle, thus, refers to advantages (returns) which might have been obtained from any factor if it had not been used in producing that commodity, but would have been used for other next-best purpose. If a cultivator for example, has Rs. 8000 for investment in his farm business, he will make a choice from various alternative uses to which this money could be put. He might have the following alternatives:

(i) Purchase of a buffalo giving net returns Rs. 800, i.e., 10% returns on the funds invested (after making the allowance of interest and depreciation and up-keep, etc.).

(ii) Purchase of a water lift on 5-acres farm; net return Rs. 960. (12% returns of investment.)

(iii) Invest on a bullock cart for transporting his produce to the market, thereby increasing share in the consumers' rupee–a net return of Rs. 600 (71/2% returns).

The farmer gets the highest return of Rs. 960 from use in water lift as compared to Rs. 800 on the purchase of a buffalo and Rs. 600 from investment on the cart. Opportunity cost of choosing one alternative' is to surrender the next best alternative. In simple terms it is the cost equivalent to the returns from next best alternative foregone.

PRINCIPLE OF COMBINING ENTERPRISES

A farm manager is often confronted with the problem as to what enterprise to select, and the level at which each enterprise should be taken up. How far he can go or should go in combining one enterprise with another enterprise or replacing one enterprise with another, depends partly on the inter-relationships between different enterprises and the prices of products and inputs.

TYPES OF PRODUCT RELATIONSHIPS

The enterprises can have any one or combinations of the following relationships:

1. Independent enterprises 2. Joint enterprises 3. Competitive enterprises 4. Supplementary enterprises 5. Complementary enterprises

1. Independent enterprises

Independent enterprises are those which have no direct bearing on each other; an increase in the level of one neither helps nor hinders the level of the other. Such a relationship is rare and is possible with practically unlimited supplies of inputs. In such cases each product should be treated separately.

2. Joint enterprises

Joint products are those which are produced together, e.g., cotton and cotton seed, beef and hides. Wheat and straw, mutton and wool, cattle and manure, etc. The quantity of one product produced decides the quantity of the other product.

In case of joint products there is no economic decision to make with respect to the combination of products and the two products can be treated as one. In the two products may necessitate a change in the product combination. For instance, a continued increase in the price of wool compared to the price of mutton may necessitate a change in the quality of the breeds which would produce a higher proportion of wool. Once this happens, again the proportions of the two products become fixed.

3. Competitive Enterprise

    (i)   The rate at which one enterprise substitutes for the other

    (ii) The prices of the products, and

    (iii) Increasing rates of substitution

CONSTANT RATE OF SUBSTITUTION BETWEEN PRODUCTS

The production possibility curve in this case would be a straight line. It means that unit change in the other product is accompanied by the opposite change in the other product at the same time throughout. For example, two crops (gram and wheat) substitute at constant rate for land, This relationship can be expressed as:

$$\frac{\Delta_1 Y_2}{\Delta_1 Y_1} = \frac{\Delta_1 Y_2}{\Delta_2 Y_1} = \frac{\Delta_1 Y_2}{\Delta_2 Y_1} = \frac{\Delta_1 Y_2}{\Delta_n Y_1}$$

For finding out an optimum combination of two products we need to work out the following :

(i) Substitution ratio $= \dfrac{\text{Amount of replaced product}}{\text{Amount of added product}} \quad \dfrac{\Delta Y_2}{\Delta Y_1}$

(ii) Price ratio $= \dfrac{\text{Price of added product}}{\text{Price of replaced product}} \quad \dfrac{P_{y1}}{P_{y2}}$

Optimum point of the combination of the two products will be located

where $\dfrac{\Delta Y_2}{\Delta Y_1} = \dfrac{P_{Y1}}{P_{Y2}}$

Rule : In this situation of constant rate of substitution, profit can be maximized by producing only one product or otherwise maximum profit will be indifferent to the combination of enterprises.

The optimum combination of two products obtained by comparing the marginal rate of substitution with the price ratio.

If marginal rate of substitution equals price ratio, any combination of the enterprises will give the same profit.

Decreasing Rate of Substitution between Products

When two products are substituting at a decreasing rate the production possibility curve is convex to the origin. It means a unit increase in the one product is accompanied by less and less decrease in the level of other product. Such conditions are seldom found in agriculture. This relationship can be expressed as under

$$\frac{\Delta_1 Y_2}{\Delta_2 Y_1} \ge \frac{\Delta_1 Y_2}{\Delta_2 Y_1} \setminus \ge \frac{\Delta_1 Y_2}{\Delta_2 Y_1} \setminus \ge \frac{\Delta_1 Y_2}{\Delta_n Y_1}$$

Rule : Under such conditions, maximum profits can be obtained by producing only one of the two products. A combination of the two products will not yield maximum profit since successive increases of one unit in the production of one product requires the sacrifice of smaller and smaller quantities of the other product. One reason for the production possibility curve to be convex is that the cost limits the use of variable input in such small quantities that both products are produced in stage I, i.e., the area of increasing marginal physical productivity.

Increasing Rates of Substitution between two Products :

Where two products are substituting at an increasing rate, the production possibility curve is concave to the origin. It means that each unit increase in level of one product is accompanied by more and more decreases in the level of another product. For example wheat and gram substitute at increasing rate for capital or labour which means when wheat acreage is increased, gram acreage will have to be decreased more and more for the purpose of releasing capital. the increasing marginal rate of substitution can be expressed as :

$$\frac{\Delta_1 Y_2}{\Delta_2 Y_1} \langle \frac{\Delta_2 Y_2}{\Delta_2 Y_1} \langle \frac{\Delta_i Y_2}{\Delta_i Y_1} \langle \frac{\Delta_n Y_2}{\Delta_n Y_1}$$

Rule : In such a situation, profits are maximized by producing a combination of the two products where substitution ratio equals the price ratio.

4. Supplementary Enterprises

Two products are said to be supplementary when an increase in the level of one does not adversely affect the production of the other but adds to the total income of the farm, i.e., enterprises which do not compete with each other but add to the total income. For example, on many small farms a small dairy enterprise, or a poultry enterprise or a small bee-keeping enterprise may be supplementary to the main crop enterprise, because they utilize surplus family labour and shelter available and perhaps even some feeds which would otherwise go to waste. In the beginning, such enterprises are added in order to fully utilize the available resources but when they are expanded far, they become competitive for the inputs.

Sometimes enterprises are supplementary for one resource but competitive for another. In such cases, the relationship should be treated as one of competitive even though they are supplementary to one another in respect of other resources(s).

5. Complementary Enterprises

Complementary enterprises are those which add to the production of each other, e.g., Berseem and Maize crops. Two products are complementary when the transfer of a variable input from the production of one product to the production of the other results in an increase in the production of both products. When two crops are complementary enterprises, the use of resources for the two crops results in the increased production of both the crops.

This relationship exists only when one enterprise produces an element which is required in the production of another enterprise, e.g., legumes and grasses become complementary to grain crops when the former (1) furnish nitrogen, (2) control soil erosion and/or (3) maintain or improve soil tilth to the extent that greater production of grain crops becomes possible over time on the same acreage. Two enterprises do not remain complementary over all possible combinations. They become competitive at some stage. When both complementary and competitive relationships occur, the complementary relationship occurs first and then is followed by the competitive relationship.

When two products are complementary, both the enterprises should be produced upto the end of the complementarily stage without reference to the prices of the two products. When they enter the competitive stage, the substitution ratio and the price ratio of the two products should be considered and the optimum combination determined as in the case of competitive enterprises.

PRINCIPLE OF COMPARATIVE ADVANTAGE

According to this well-known principle, different areas will tend to produce those products for which they have the greatest comparative, and not just absolute, advantage. This leads to the establishment of different types of farming existing in a particular area.

TIME COMPARISON PRINCIPLE

Farm management involves dynamic adjustment of the farm organization and operations. Such an adjustment relates to taking account of: (a) time element in the calculation of present value of future incomes and (b) the risks and uncertainties involved in farm operations over time. The first adjustment involves a method of discounting the future returns. The rate of discounting will perhaps be lower for a farmer with unlimited capital than for a farmer with limited capital.

The risk and uncertainties are a general rule in agriculture. While some risks are insurable, the others are not. The most important risks and uncertainties occur due to natural calamities, fluctuations in prices of inputs and outputs and changes brought about by innovations and other technological improvements. Producers tend to discount returns to adjust for these uncertainties. Some farmers are security minded and are adverse to taking risks. Such farmers will be reluctant to adopt new practices or enterprises. Thus, they discount for the uncertainties at a very high rate.

Time has a very significant influence on costs and returns and as a result involves considerations which should be recognized in managerial decisions. There are many decision situations where this principle finds application, such as: soil conservation programmes which bear fruit over a long time; use of land for growing crops or putting it under an orchard which may not give returns for 5 to 10 years; purchasing of dairy cows in milk by paying more or purchasing heifers and raising them till they begin to produce.

GROWTH OF A CASH OUTLAY

The cash outlay grows over time due to the compounding of interest charges or opportunity costs involved in using the capital : If Rs. 100 are put in a saving-account with annual interest at % compounded, it will increase or grow to Rs. 127.63 by the end of five years. A cash outlay or investment made in a farm business grows over time in a similar manner. Farm improvements, machinery, breeding stock, milk cows are examples. While such items depreciate over time, growth of the remaining capital invested has to be considered in the analysis whether the investment will be profitable. Suppose, for example, it costs Rs. 30 per acre to grow nepier grass which would be in production for three years, with one-third of the capital outlay charged off each year. Thus, one-third of the Rs. 30 (Rs. 10) would be invested for one year, another Rs. 10 for two years and Rs. 10 for three years. The value of this investment would grow to Rs. 33.10 by the end of the three-year period. Thus, Rs. 33.10 would be appropriate cost to use in determining whether it would pay to plant nepier grass.

Since costs grow as a result of interest or opportunity cost accumulations, the equation for compounding interest may be used to show growth in the cash outlay:

$S = s(1+i)$ 'S' represents the sum at the end of 'n' periods; 's' the amount which is invested for 'n' periods; 'i' the interest rate.

Discounting Income

Discounting income is the procedure whereby the present value of the future income is determined. The concept is the reverse of "growth in value" due to accrued interest. Thus with interest rate at 5% Rs1.00 today grows to Rs. 1.05 in a year and conversely, Rs. 1.05 a year from now is worth only Rs. 1.00 today.

$$PV = \frac{q}{(1+r)^n}$$

PV = present value of the future amount

q = future amount

r = rate of interest

n = No. of years in the future.

Despite the above limitations, farm management has an important role to play under Indian conditions. Even though majority of the farmers produce mainly for subsistence, they can aim at maximization of the gross production rather than of net income.

With liberal economy the farmers would become more and more commercial minded and they would be able to effect technological and economic adjustments aimed at stepping up production and net income. Farm management principles are, therefore, likely to gain grater implicational value in the near future.

Farm Layout Planning Budgeting and Resource Management

Layout of farm refers to the manner in which a farm is divided into fields and the location and arrangement of other fixtures such as irrigation and drainage system, buildings and sheds, roads, fencing, etc. Layout of a farm directly affects:

1. Costs and efficiency in the use of man power, bullock power and machinery,

2. Costs and efficiency of irrigation, drainage and fencing, and

3. Cropping plan and profitability of the farm business as a whole.

PLANNING THE FARM LAYOUT

The following factors offer a general frame work for a good farm-layout

(i) The type of the farm : experimental, demonstration, seed multiplication, or commercial (Vegetables, orchards, general crop farming, dairy, etc.).

(ii) The general topography of the farm land along with other environmental factors.

(iii) Variation in soil fertility.

(iv) Irrigation and drainage problem.

(v) Size of the farm.

(vi) Cropping scheme.

(vii) Livestock kept.

(viii) Building structure.

(ix) Farm communication situation-roads and field paths.

Consistent with the size and purpose of the farm topography and system of farming, a good farm layout should ensure : A small number of good-sized fields, Rectangular fields, Minimum area under buildings, roads and water channels, Buildings located in the centre of the farm, Easy access to every field, Minimum fencing cost, Uniformity of the soil within a field and within major blocks and Efficient and economical irrigation and drainage system.

An efficient layout is the one which takes into consideration the topography of the land, fits in well with the enterprises and crop rotations, leads to the saving of labour and ensures efficient checks and controls on farming operations.

It is usually not easy to make changes in the farm layout because it involves investment. But with the changing techniques of production and increasing mechanization, remodeling of the farm layout become necessary.

Mechanisation-replacing bullock power with tractors. Changing the source of irrigation - replacing well irrigation with canal irrigation.

Soil Conservation – Controlling the Soil Erosion

There can be many other reasons that will compel a farmer to change the layout of farm.

Thus, the layout of the farm has a strong bearing on the organization and allocation of the resources to different enterprises (*i.e.,* the crop pattern, and determines the profitability of the farm business as a whole.

Farm Planning : Most farmers probably have some kinds of plans about the organization and operation of their farms. Perhaps; in most cases, they are not very systematically worked out, they may be based largely on habit or customs.

Definition : A farm plan is a programme of the total farm activity of a farmer drawn up in advance. "The improved farm planning is a process of observation, appraisal and analysis of weighing the merits of new and old ideas and then deciding which ideas to use in the period ahead" - Malon, Carl C. This process is learned much better if the results of the thinking and reasoning are written down in a simple but organized way. Hence various farm planning proforma (written forms) are used for preparing farm plans.

Uses of Farm Plan : The major use of a farm plan prior to the beginning of a season is to outline the programme of work, study the same indicating planned organization and operational practices from the view point of management principles. A farm plan is best used during the operations as a flexible guide and not as a fixed rule for every operation. A farm plan serves as a compass to keep the farm manager on the right track. The comparison of the estimated results with actual results from each of the individual enterprises and the entire farm business serve to make future plans more effective by locating strong and weak points of the farm business. Farm planning has to be based on the actual conditions of the farms concerned. The length of the planning period on the basis of the farmers situation has to be decided to ensure maximization of the objectives envisaged. The main objective is to obtain maximum the annual net income sustained over a long period of time. The ultimate objective of farm planning is the improvement in the standard of living of the farmer. By working out several alternative plans the farmer can of course, make his choice on the basis of several objectives. The advantages of farm planning approach lies in its treating the farm as an operational unit and tailoring the recommendations to fit into the individual farmers' opportunities, limitations, problems and resource position.

PRINCIPAL CHARACTERISTICS OF A GOOD FARM PLAN

Under the Indian conditions a typical farm plan should have the following characteristics and information. It should provide for efficient use of farm resources such as, labour, power and equipment. The crop plan should have balanced combinations of enterprise *i.e.*, it should provide for given minimum production of different food, cash and fodder crop, Help maintain and improve soil fertility, Help raise and stabilize farm earnings and requirements throughout the year. Avoid excessive risks. Provide flexibility. Utilize the farmer's knowledge, training and experience and his likes and dislikes. Give considerations to efficient marketing facilities. Provide programme of obtaining, using and repaying the credit. Provide for the use of up-to-date modern agricultural methods and practices.

INFORMATION NEEDED FOR PLANNING AND BUDGETING

(a) Farm planning under perfect knowledge. The information needed for farm planning can be grouped into five categories

1. Resources available on the farm statement of resource restrictions.

   (i) Topography, (ii) Drainage (iii) Soil management problem.

   (i) Land holding (ii) Irrigation potential, (iii) Labour on the farm, (iv) Working capital (v) Farm buildings

   Additional resources that can be acquired.

2. Outputs to be produced- List of process and farm enterprises.

3. Technical input-output coefficients.

4. Expected prices of farm products and inputs.

5. Social, institutional and personal framework within which the farmer operates his farm business.

(b) Farm planning under imperfect knowledge :

   Additional data required under imperfect knowledge are :

   (i) Price and yield variability information.

   (ii) Time series data on prices.

   (iii) Information on probability distribution for probabilistic game theory models.

   (a) Possible sets of events - weather conditions, credit policies, various price levels.

   (b) Alternatives available to the farmer.

   (c) Specification of consequences - in monetary terms or in utility terms.

   (d) Subjective probability distribution attached to a set of events.

FARM BUDGETING

Farm budgeting is a method of analysing plans for the use of agricultural resources at the command of the farmer decision-maker.

The expression of a farm plan in monetary terms by estimation of receipts, expenses and net income, is called budgeting. In other words, farm budgeting is a process of estimating costs, returns and net profit of a farm or a particular enterprise.

The farm budget helps the farmer to evaluate alternative plans and select the one that is most suitable. Planning without calculating the returns and costs in monetary terms, is not of much use. So farm planning and budgeting go side by side.

Types of farm budgeting : There are two methods of farm budgeting. (a) Partial budgeting (b) Complete budgeting.

(a) Partial budgeting

It refers to estimating the outcome or returns for a part of the business, i.e., one or a few activities. It provides a method for deciding low expenses and yields should be increased of a particular enterprise. Changes in organization can be worked out without complete working of the whole plan. This change may be made through a careful selection of alternative methods of production or practices or activities, the choice of which is based on opportunity cost or relative profitability and does not affect the total farm organization vitally. Partial budget fails to consider all the relevant factors in maximizing net returns to the farm as a whole. It also does not allow substitution between resources and it over looks complementarily and competition between different enterprises.

Format of partial budget will look like this

Debit	*Credit*
(a) Increase in costs	Decrease in costs
(b) Decrease in returns	Increase in returns
Net gain	Net loss

ENTERPRISE BUDGET

Enterprise budget is used to estimate inputs required, costs involved, and expected returns from a particular enterprises. The purpose of budgeting an enterprise to aid in selection of inputs and enterprise consistent with the resources available and to show combination(s) that will increase income from the farm business.

The enterprise budget includes only variable costs while the cost of cultivation includes both fixed and variable costs.

The enterprise budget prepared for the practices followed by the farmer when compared with the one at the recommended package of practices provides the straight-forward information about the potentials that can be harnessed. Several kinds of data are necessary for budgeting an enterprise, physical inputs data (seed, fertilizer, insecticides, etc.) field output data (yields per hectare at different levels of inputs) Price data for inputs and outputs. Selection of data to use in enterprise budget requires considerable judgment.

(b) Complete Budgeting

It refers to making out a plan for the whole farm organization or for all decisions on one enterprise. Complete budgeting, considers all crops, livestock producing methods and estimates costs and returns for the total farm. Thus, it presents a complete picture of the farming business. It indicates the worth of the resources in terms of their prospective earning capacity.

Farmers need to make both long run and short run or annual budget plans. It is advisable to prepare a tentative plan on a long term basis and phase the programme into annual plans. Short term or Annual plans - Emphasis on operational improvement. Long term plan considers structural improvements which are of a permanent nature.

Complete budgeting draws attention to the variety of factors affecting farm income. It considers complementary, supplementary and competitive relationships between enterprises. It allows the substitution between resources and avoid omission of any vital part of the organization being left out of consideration.

STEPS IN COMPLETE FARM PLANNING AND BUDGETING

1. Appraisal of the existing farm resources, their use pattern and efficiency, *i.e.*, the constraining framework.

2. Appraisal of the alternatives or opportunities, i.e., various production activities that can be included on the farm organization and their resource requirements (input-output coefficients).

3. Preparing and evaluating the alternative plans for their feasibility and profitability.

   Thus complete budgeting requires more time, efforts and more basic data in accurate farm.

LABOUR MANAGEMENT

Labour efficiency in agriculture refers to the amount of productive work. accomplished per man on the farm per unit of time. In general, the higher the efficiency, the greater are the returns from farming.

On Indian farms, land is limited, *i.e.*, farms are small in size, capital is extremely limited and organizational or managerial ability is low. Labour, on the other is abundant. Thus the resource availability on the farm is imbalanced leading to a low production which results in low returns to the farm business and low farm family labour earnings. The situation thus creates a vicious circle. This vicious circle does not allow the farm labour a living wage and poverty conditions persist in the farm sector, especially among the agricultural labour.

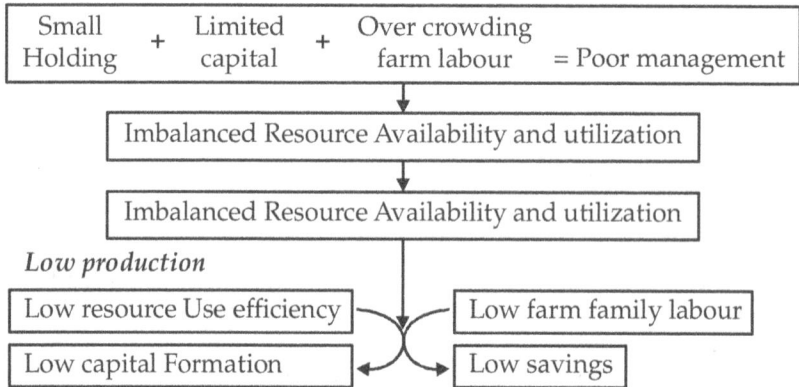

Small Holding	+	Limited capital	+	Over crowding farm labour	= Poor management

↓

| Imbalanced Resource Availability and utilization |

↓

| Imbalanced Resource Availability and utilization |

Low production

| Low resource Use efficiency | | Low farm family labour |

| Low capital Formation | | Low savings |

The labour as a basic farm resource has important characteristics, some of which are peculiar to farm labour only. These are : They are large in number. They are mostly unskilled. They lack staying power. Majority of farm labour are under debt and lack bargaining powers. Payment of wages are low, many times seasonal and in kind. Their hours of work are long and irregular and employment opportunities are meager, seasonal and uncertain.

Classification of farm labour : Farm labour can be classified into 4 categories (i) Farm mange's labour (ii) Farm family's labour (iii) Permanently hired labour and (iv) Causal hired labour. The first three categories constitute permanent labour force available on the farm and is a fixed resource due to a general lack of mobility. Casual hired labour is a variable input and can be hired when needed.

Improving the efficiency of farm labour : To the farmer, increasing the efficiency of labour especially when it is hired is an important consideration. Some of the methods which have been found to be useful in improving the labour efficiency are :

1. Enlarging the size of the farm business,

2. Planning labour distribution and enterprise combination

3. Improving the field and the farm layout,

4. Improving the farm buildings programme,

5. Improving labour management with planning of the work, incentives, and training,

6. Farm work simplification to enable the worker show his skills in effective manner.

THE ROLE OF MANAGEMENT

The farm is confronted with an imperfect knowledge situation on one or the other count but he has to reach a definite decision which is going to implement and for this he has to make predictions about the future and farm expectations. Then he has to make a plan evaluating the alternatives which are consistent with his expectations. Action has to be taken on the final decision, i.e., the decided plan is to be put into effect. And finally, the management bears the responsibility, i.e. stands the consequences of the outcome, of which may be good, or bad. It is the prediction of the consequences to be good or bad and the farmers ability to withstand the range towards bad consequences that varies from farmer to farmer and explains the deviations in the farmers actual plan from the one which could have been restored under perfect knowledge. Thus, it is the farmers ability to learn analyse, and predict more efficiently that matters in his ultimate success in the farming business.

PLACE OF AGRICULTURE IN INDIAN ECONOMY

Agriculture plays key role in Indian economy as contribute major share of national income, largest employment providing sector, supply raw material to industry, markets for industrial products and share in international trade.

Share in national income : At the time of first world war, agriculture contributed two third of national income. This was on account of non-existence of industrial development. The share of agriculture in national income during post independence period declined from 59.2 per cent in 1950-51 to 34.9 per cent in 1990-91 and 27.5 per cent in 1999-2000.

Contribution to Employment : In 1951 the working population engaged in agriculture was 69.5 per cent fell to 59 per cent in 2001. However, with rapid increase in population the absolute number of people engaged in agriculture has become exceedingly large. This indicates need for development of agro-industries and other sector to provide gainful employment.

Source of raw material to industry : Agriculture provides raw materials to various industries of national importance. Sugar industry, jute industry cotton textile industry, oil and vanaspati are major industries depend on agriculture for their development. The entire range of food processing industries depends on agriculture. Therefore, unless agriculture develops, these industries will also remain backward.

Market for industrial products : Increased rural purchasing power is a valuable stimulus to industrial development. Some of the inputs like fertilizer and chemicals are used in agriculture.

Importance in international trade : The cotton textile, jute and tea accounted for more than 50 per cent of export earnings of the country. Such heavy dependence on agriculture commodities for export earning reflected underdeveloped nature of the economy. The share of agricultural exports in total exports was 44.2 per cent in 1960-61 fell to 30.7 per cent in 1980-81 and to 15.1 per cent in 1999-2000 and 11.20 in 2005-06 inspite of absolute increase in multiple.

Employment income and export of India

Year	Population million	Agril. Employment million	Share of agril. In NDP	Export Rs. crore		Share of agril. exports
				Total	Agril	Agril. In Export
1960-61	439.2	131.1	54.0	642	281	44.24
1970-71	548.2	125.7	47.4	1535	487	31.73
1980-81	683.3	148.0	36.5	6711	2057	30.65
1990-91	846.4	185.3	34.9	32527	6013	19.40
2000-01	1028.7	234.1	24.6	201356	28657	14.23
2004-05	1091.0	215.3	22.3	356069	39863	11.20

AGRICULTURAL PRODUCTION AND PRODUCTIVITY TRENDS

For assessing the performance of the Indian agricultural sector, it is necessary to discuss the production and productivity trends in agriculture.

TRENDS IN AGRICULTURAL PRODUCTION AND PRODUCTIVITY

Agricultural production has two components - foodgrains and non-foodgrains. The former contributes approximately two thirds of total agricultural production. Trends in agricultural production and productivity are presented in tables respectively. Let us consider table first. As far as foodgrains output in concerned, the total production increased from 50.8 million tonnes in 1950-51 to 155 million tonnes in the Seventh Plan (annual average) and further to 187 million tonnes in the Eighth Plan (annual average) Production of foodgrains touched the record level (highest in the post-Independence period) of 209 million tonnes in 1999–2000 and 231 million tonnes in 2007–08 .

For purpose of analysis the entire table can be conveniently divided into two parts (i) period up to the end of the Third Plan and (ii) period after the Third Plan. The latter is often referred to as the period of the Green Revolution and as would be noticed from table is marked by rapid strides in wheat with jowar, bajra and maize continuing to show erratic trends as in period (I) Excepting setbacks in some years, rice shows a steady upward trend.

In the non-foodgrains group, jute and cotton show slow and halting progress in both the periods. However the last few years have been very good as far as oilseeds production is concerned. In fact oilseeds registered a phenomenal increase in production from 12.7 million tonnes in 1987-88 to 18.0 million tonnes in 1988-89 and further to 18.6 million tonnes in 1990-91. The production of oilseeds in the Eighth Plan rose to 21.9 million tonnes (annual average and further to a record level of 24.7 million tonnes in 1998-99). (However, the production of oilseeds fell to 20.9 million tonnes in 1999-2000). Because of the substantial increases recorded by the production of oilseeds in recent years, some economists have argued that the Indian agriculture has witnessed not just one crop revolution but two crp revolutions a Green Revolution and a Yellow Revolution. Production of cotton rose from 8.4 million bales in the Seventh Plan (annual average) to 12.2 million bales in the Eighth Plan (annual average). During 1999-2000 the production of cotton was 11.6 million bales. Sugarcane registered a more or less steady growth during the entire period 1950-51 to as high as 2,755 kgs. per hectare in 1999-2000. Productivity of coarse cereals jowar, bajra and maize has risen relatively slowly. Most disappointing has been the performance of pulses. In fact, productivity of pulses was only slightly higher in 1999-2000 as compared with 1960-61.

Yield per Hectare of Major Crops

(kgs per hectare)

Crop	1950-51	1960-61	1971-72	1985-86	1998-99	1999-2000	2005-06
Rice	668	1,013	1,141	1,552	1747	1990	2102
Wheat	655	851	1,380	2,046	2590	2755	2619
Jowar	353	533	460	633	859	852	880
Bajra	288	286	452	344	748	639	802
Maize	547	926	900	1,146	1797	1785	1938
Pulses	441	539	501	547	634	630	598
Total Foodgrains	552	710	858	1,175	1571	1697	1715
Oilseeds*	481	507	546	570	944	856	1004
Cotton	88	125	151	197	224	226	362
Jute	1,044	1,183	1,255	1710	1825	1995	2173

* includes groundnuts, rapeseed and mustard, seasamum, linseed and castor seed.

LOW LEVELS OF PRODUCTIVITY

Productivity is generally considered from two angles - (i) productivity of land and (ii) productivity of labour engaged in agriculture table gives information on the former. As is clear from this table, there has been a slow and steady rise in productivity during 1950-51 to 2005-06. For most of the crops GDP per hectare was Rs. 875 in 1950-51 which rose to Rs.1,023 in 1960-61 Rs. 1,204 in 1970-71 and further to Rs. 1.401 in 1979-80 (at 1970-71 prices). Thus output per hectare rose by 60 per cent over the period 1950-51 to 1979-80. However, productivity per worker has remained almost stagnant over the period as would be clear form the fact that GDP per worker which was Rs. 1,019 in 1950-51 fell to Rs. 988 in 1960-61, rose to Rs. 1,013 in 1970-71 and to Rs. 1,025 in 1979-80 (at 1970-71 prices).

A comparison of productivity levels in Indian agriculture with the levels in other countries shows how the productivity is low in India. Table indicates, productivity of wheat in India is about 34 per cent of the productivity in France. It is 70 per cent of the productivity in comparison to another developing country, China. As far as rice is concerned, productivity in India is 48 per cent of the productivity in China and 46 per cent of the productivity in Japan (i.e., less than half). The productivity of cotton in India is only about one-third of the productivity in China. Even in comparison to Pakistan, productivity of cotton in India is just 57 per cent. As far as groundnut is concerned, productivity in India is 34 per cent of the productivity in USA, 40 per cent of the productivity in China and 43 per cent of the productivity in Argentina. Similar conclusions hold for most of the other crops not included in the table. The low levels of productivity in Indian agriculture point to the possibilities of increasing productivity by adopting right strategy and intensive efforts.

Productivity of Land in Some Countries, 2006. (100 kgs /ha)

Crop	Productivity	Crop	Productivity
Wheat		*Cotton (Lint)*	
U.K.	80.4	China	8.4
France	67.4	U.S.A.	6.9
China	44.5	Pakistan	5.3
U.S.A.	28.5	India	3.0
India	26.2	Groundnut (in shell)	
Rice (Paddy)		U.S.A.	29.6
Egypt	105.9	China	31.2
Japan	63.4	Argentina	30.3
China	62.6	India	8.59
Indonesia	47.7		
India	31.2		

CAUSES OF LOW PRODUCTIVITY

As shown above, though agricultural productivity has increased in India, it is still very low when compared with other countries. The causes of low productivity can be divided into the following three categories : (i) general, (ii) institutional, (iii) technical.

GENERAL CAUSES

Social environment : The social environment of villages is often stated to be an obstacle in agricultural development. It is said that the Indian farmer was illiterate, superstitious, conservative and unresponsive to new agricultural techniques. On the face of it, this seems to be correct. However, the fact is that given the limitation of present production relations, the unassuming and ignorant looking farmer uses his resources efficiently.

Pressure of population on land : The pressure of population on land is continuously increasing. Whereas the number of people dependent on agriculture was 16.3 crore in 1901, it rose to 44.2 crore in 1981. Though, additional land has been brought under cultivation since 1901, yet per capita cultivated land has declined from 0.444 ha. in 1921 to 0.296 hectares in 1961 further declined to 0.219 ha. in 1991and 0.19 ha. in 2001. Increasing pressure of population on land is partly responsible for the sub-division and fragmentation of holdings. Productivity on small uneconomic holdings is low.

INSTITUTIONAL CAUSES

Land tenure system : Perhaps the most important reason of low agricultural productivity has been the zamindari system. Highly exploitative in character, this system drained out the very capacity, willingness and enthusiasm of the cultivators to increase production and productivity. Legislation's passed for abolition of intermediaries in the post-Independence period did not break the stranglehold of the zamindars on the rural economy. They only changed their garb and became large landowners. Exploitative practices continued. Regulation of rent, security of tenure, ownership rights for tenants, etc., did not make the position of tenants better. Tenancy of most of the tenants continues to be insecure and they have to pay exorbitant rates of rent. In this land tenure system, it is difficult to increase productivity only through technological means. In fact, land reforms should precede technological changes. If investment in agriculture has to be increased it is necessary to eliminate the renter class of zamindars and spurious class of moneylenders.

Lack of credit and marketing facilities : It is often assumed that the decisions of Indian farmers are not affected or modified in response to price incentives. In other words, the Indian farmer continues to produce the same agricultural output even on more attractive prices. However, the facts are different. Frequently on account of lack of marketing facilities or non-availability of loan on fair rate of interest, the cultivators are not able to invest the requisite resources in agriculture. This keeps the level of productivity on land and per cultivator low. If the government can revitalize the credit cooperative societies and the regional rural banks to grant more credit to the small farmers, the level of productivity can undoubtedly increase.

Uneconomic holdings : According to the National Sample Survey, 52 per cent hloldings in 1961-62 had a size of less than 2 hectares. In 1990-91, 78 per cent of total holdings fell under this category. Most of these holdings are not only extremely small they are also fragmented into a number of tiny plots so that cultivation on them can be carried out only by labour intensive techniques. This results in low productivity. Until the excessive labour employed on agriculture is transferred to alternative jobs and the holdings are consolidated (or cooperative farming initiated) modern techniques of agriculture cannot be adopted and the possibilities of increasing agricultural productivity will remain limited.

TECHNICAL CAUSES

Outmoded agricultural techniques : Most of the Indian farmers continue to use outmoded agricultural techniques. Wooden ploughs and bullocks are still used by majority of farmers. Use of fertilizers and new high-yielding varieties of seeds is also extremely limited. In summary, Indian agriculture is traditional. Therefore productivity is low.

Inadequate irrigation facilities : Gross cropped area in India in 2001 was 187.94 million hectares of which 75.14 million hectares had irrigation facilities. Thus, 39.98 per cent of gross cropped area had irrigation facilities in 2001. This shows that even now about 60 per cent of the gross cropped area continues to depend on rains. Rainfall is often insufficient, uncertain and irregular. Accordingly, productivity is bound to be low in all those areas which lack irrigation facilities, and are totally dependent on rains. Even in areas having irrigation facilities, potential is not wholly utilized because of defective management. The costs of irrigation are also increasing continuously and the small farmer is, therefore, unable to make use of available irrigation facilities.

MEASURES TO INCREASE PRODUCTIVITY

The causes given above also suggest the measures to increase productivity. As would be clear, such measures would have to attack the problem from technical, institutional, social and economic angles. In particular, attempts will have to be made in the following directions.

Implementation of land reforms : Though land reforms have been introduced in India in the post-Independence period with a view to eliminating the intermediary interests in land (especially zamindari), providing security of tenure and ownership rights to tenants, and reorganizing agriculture through land ceiling legislation, cooperative movement and consolidation of holdings, the progress registered is too unsatisfactory. Therefore special attempts will have to be made by the State governments to implement the land reforms legislation carefully so that the slogan 'land to the tiller' is translated into practice. Unless this is done, the tiller will have no incentive to invest in land and adopt new agricultural techniques. Therefore, land reforms are the first and foremost necessity.

Integrated management of land and water resources : The total geographical area of the country for which information is available is 304.9 million hectares of which only 264.0 million hectares posses potential for biotic production. Of this, wastelands account for 79.5 million hectare leaving only 184.5 million hectares. However, even this area cannot be regarded as being in good health. According to the land use statistics, the total extent of lands that suffer from degradation, to a greater or less degree, is 175 million hectares. Since this figure obviously includes wastelands, it follows that the area of lands that are still productive but are suffering from degradation is 95.5 million hectares (175 million hectares minus 79.5 million hectares). Since this area of 95.5 million hectares must necessarily be a part of the 142.2 million hectares of land that is under agriculture, it means that nearly two-third of our agricultural lands are sick to some extent or another. This is quite alarming. In fact, as pointed out by B.B. Vohra, of the nearly two third of our total land resources, suffering from degradation, almost 50 per cent have undergone such degradation that they have, for all purposes, ceased to be productive. This proves the urgency of an integrated and efficient management of our land and water resources. It is particularly important to control soil erosion which affects around 150 million hectares out of the country's total land area of 304.9 million hectares as it constitutes the biggest single threat to the sustainability of our agriculture.

Improved seeds : Improved seeds can play an important role in increasing productivity. This has been amply proved by the experience of many countries and by the demonstration of high-yielding varieties of wheat in Punjab, Haryana and Uttar Pradesh in our own country. Therefore more and more farmers in more and more areas should be encouraged to use improved seeds. After examining the soil conditions and availability of irrigation facilities in different areas, farmers should be advised about which seeds are best in the area. They should also be educated in the methods of sowing, manuring, and irrigating the new high yielding varieties of seeds.

Fertilizers, Improved varieties of seeds require heavy doses of fertilizers. It has been estimated by agricultural scientists that Indian farmers use only one tenth the amount of manure that is necessary to maintain the productivity of soil. Fertilizer used per hectare was merely 95.3 kg in India against 262.3 kgs in China, 130.1 kgs in Bangladesh and 306.2 kgs in Egypt in 1997-98. It has been estimated that use of fertilizers in ample quantity (especially nitrogen, phosphorus and potassium) can push up the productivity manifold.

Irrigation : Use of improved seeds and fertilizers requires proper irrigation facilities. Irrigation can also make multiple cropping possible in a number of areas and hence enhance productivity. Attempts in this field will have to be undertaken in the following directions modernizing irrigation systems in a phased manner, better operation of existing systems, efficient water management, adequate maintenance of canals and distribution systems, detailed surveys and investigations for preparation of new projects, developing a National Grid System to ensure water supply from water surplus areas to water deficit areas, etc.

Plant protection : Agricultural scientists have estimated that approximately 5 per cent of the crops are damaged by insects, pests and diseases. Most of the farmers in the countryside are unaware of the medicines and insecticides developed in recent years to face this challenge posed by diseases and insects. Some farmers have started using them to some extent but their efforts cannot be successful unless and until their neighboring farmers also adopt them. Therefore, it is necessary to manage this programme at the government level. The government should maintain its own technical staff to carry out the spraying of pesticides and insecticides at nominal rates.

Farm mechanization : It is generally believed that through farm mechanization agricultural productivity can be increased. Supporters of mechanization argue that it results in increase in productivity of land and labour, reduction of costs, saving of time and increase in economic surplus. However, it should be borne in mind that all estimates of productivity include the contribution of machines as well as other agricultural inputs like improved seeds, fertilizers, etc. and it is not possible to say how much of increase in productivity is due to mechanization alone. Nonetheless, it cannot be denied that mechanization saves labour time which can be utilized elsewhere.

Provision of credit and marketing facilities : Use of improved varieties of seeds, fertilizers, pesticides, insecticides, agricultural machinery and irrigation facilities all require substantial money resources which small farmers do not usually possess. Therefore, it is necessary to strengthen the credit cooperative sector and free it from the clutches of large landowners so that it can meet the credit requirements of small farmers. The commercial banks should be encouraged to expand credit to small farmers. Regional rural banks can play a special role in this regard. The marketing structure also needs a reorientation to serve the small and marginal farmers in a better way. Cooperative marketing societies should be promoted to ensure better prices to small farmers.

Incentives to the producer Incentives to the agriculturists can go a long way in encouraging them to increase productivity. Incentives can be in the following forms: (a) implementing land reforms rigorously and vigorously, (b) ensuring timely availability of agricultural inputs, (c) guaranteeing remunerative prices of produce to the farmer (d) implementing crop insurance scheme to cover the risk of damage to crops and other risks in agriculture, and (e) social recognition and conferment of awards, merit certificates, etc.

Better management Just as industry needs skilled management for increased productivity, agriculture also requires better management for raising the level of productivity. For this purpose farmers have to be educated in more efficient use of their resources particularly land, irrigation facilities and agricultural implements. A related problem is the extension of science and technology in agriculture. This can be accomplished only if there is a vast network of managerial staff engaged in disseminating information about new agricultural techniques and methods of production. Other tasks of this extension staff could be to test the suitability of soil and climatic conditions for different crops and advising the farmers accordingly, ensuring proper warehousing and marketing facilities, arranging for timely supply of agricultural inputs, and advising farmers on day-to-day problems confronted by them in carrying out agricultural activities.

Agricultural research : Agricultural research is presently being conducted by the Indian Council of Agricultural Research, various Agricultural Universities and other institutions for evolving high yielding varieties of seeds for different crops. Considerable success has been achieved in the case of wheat. However, intensive efforts are required for achieving similar success in other crops. Research should also be conducted on a substantial scale at different regional centers for testing the quality of soil suggesting measures for soil conservation and reclamation, examining the diseases affecting different crops, improving the quality of agricultural implements, avoiding wastage in agriculture especially damage to crop resulting from pests, insects, rodents, etc.

POTENTIAL OF PRODUCTION

As shown in table earlier, agricultural productivity in India is very low as compared to other countries. By adopting proper strategies it should be possible to push up the levels of productivity considerably. High yielding varieties of various crops have been evolved which show significantly larger yield per hectare in farm demonstration plots while under actual cultivation conditions, the productivity has been far less. For instance, even in he case of wheat (the success crop of green Revolution), the actual yield in 1999-2000 was only 2,755 kgs, per hectare as against the potential of 6000/6800 kgs. Per hectare. The gap between potential and actual productivity of various crops is given in *table*.

Potential and Actual Productivity

(kgs. Per hectare)

S. No.	Crop	Gap	Potential	Actual (1999-2000)
1.	Rice	4000/5810	1990	2010
2.	Wheat	6000/6800	2755	3245
3.	Jowar	3000/4200	852	2148
4.	Maize	6000/8000	1785	4215
5.	Groundnut	2000/3000	774	1226
6.	Cotton	700/850	226	474
7.	Jute	2500/3000	1995	505
8.	Sugarcane	96000/112000	71000	25000

INSTITUTIONAL RESTRUCTURING

It is clear that no one uniform extension system will serve as a remedy to all States. Even within States there will be a combination of various agencies and different institutional arrangements to address needs of differing agroclimatic zones as well as different sections of farmers. A menu of various models will be available to the States to select and adapt to their own requirements.

RESTRUCTURING PUBLIC EXTENSION

It will continue to remain central to technology dissemination, small and marginal farmers and economically backward regions will need to be serviced by it. This implies that the 1,00,000 public extension functionaries (including VEWs and SMSs) will have to be placed in new decentralized institutional arrangements which are demand driven, farmer accountable, bottom up and have a farming Systems. Approach (broad based). States have before them several models. Namely, (i) the ATMA model (6 States) (ii) Single Window-broad based extension Model (Maharashtra), Panchayati Raj Institutions (Kerala, W.B., M.P.) and the SAU Farmer Direct Contact (Punjab).

Agricultural Technology Management Agency (ATMA) Model

A key concept is to decentralise decision-making to the district level through the creation of the ATMA as registered society. A second goal is to increase farmer input into programme planning and resource allocation, especially at the block level and to increase accountability to stakeholders. A third major goal is to increase programme coordination and integration between departments so that the following programme thrusts can be more effectively and efficiently implemented.

(i)   **Farming System Innovations :** Especially the intensification and / or diversification into high value commodities and / or value added marketing and processing activities,

(ii) **Farmer Organizations :** Especially for high value commodities and resource poor farmers.

(iii) **Technology Gaps :** In both crop and livestock production systems, and

(iv) **Natural Resource Management :** especially soil and water management and to reduce pesticide use through integrated pest management (IPM) programmes.

AGRICULTURAL FINANCE

Credit plays a very important role in the agricultural management due to the prevalent uncertainties, low returns, high rate of rent etc. The requirements of credit may be long term, medium term or short term depending on the purpose for which the credit is required, i.e., it may be for development to make investments on the farm which include purchase of land, farm machinery etc., development of irrigation systems; land reclamation and developmental work, development of dairy, poultry, fishery, sericulture etc. or for the production to purchase seeds, fertilizers, manures, pesticides etc. and finally credit required for marketing. All these credit requirements are met by various institutional agencies like the Government, Commercial Banks, Regional Rural Banks or the highly developed co-operative network. The credit structure is as follows

CREDIT AND TYPES OF CREDIT (ACCORDING TO PURPOSE, PERIOD AND SECURITY)

Credit is power to use someone else's funds in exchange for promise to pay at a later day alongwith interest. It means trust in borrowers ability to repay and willingness to repay.

Agricultural credits may be classified on the basis of (a) the purpose for which it needs (b) the length of the period for which loan required (c) the security against which loans are advanced.

(a) According to purpose

(1) **Agricultural :** Credit needed for purchase of seed, manure, payment of rent, wages, revenue, cess and other charges.

(2) **Non- farm business purpose :** Needed for repair of production and transport equipment and furniture, current expenditure in non-farm business purchase, repair, construction of building or capital expenditure on non-farm.

(3) **For family expenditure :** Needed for purchase of domestic utensils and clothing, paying for medical educational and other family expenses.

(b) According to length of period

(1) **Short term credit :** Repayment period for 15 months to meet current expenses of cultivation, to facilitate production and meeting domestic expenses.

(2) **Medium term loan :** Repayment period 15 months to 5 years required for the purpose of making some improvements on land, buying cattle, agricultural implements, gardening, fencing, plantation etc. purchase of share of cooperative sugar factories etc.

(3) **Long term loan :** For the purpose of buying additional land, make permanent improvement on land, purchase of tractor, oil engine, machinery. These loans have repayment period from 15 to 20 years period are required.

(c) **According to security**

(1) **Farm mortgage credit :** Which is secured against land by means of mortgage of land.

(2) **Chattel or collateral credit :** The farmer is given on the basis of farmers livestock, crops or warehouse receipts, property like shares, bonds or insurance policies.

(3) **Personal credit :** Which is advanced on the promissory or personal notes of the farmer with or without another security.

Economic feasibility tests of credit : Bank assesses **Three R's of credit-** Returns, Repayment capacity and Risk bearing ability :

1. Returns : Returns from investment is an important measure in the credit analysis. The demand for credit can be accepted only when the returns estimated exceed the investment. To estimate the additional costs to be made good by the investor the partial budgeting technique is applied.

2. Repayment capacity-It means the ability of the farmer to clear off the loan obtained for business within the time stipulated by the bank. The loan amount may be productive enough to generate additional income to the borrower as well as repay the loan. In other words, the loan should be profitable and also have potential for effecting repayment. The repayment capacity depends on difference between working expenses and gross returns, family requirements, other loans due, management skill .

Causes of poor repayment capacity include small size of holdings, low productivity, low prices during post-harvest periods, low income, misuse of loan, high family expenditure, low farmers equity and poor management.

Measures to strengthen the repayment capacity - organization and operation of farm to increase net income, technology for increasing production at reduced costs, scheduling loan repayment as per the income flow, adopting risk management strategies like crop insurance, hedging to control price variations etc.

3. Risk bearing ability : It is ability of the farmer to withstand the risks that arise due to financial loss.

Three C's of credit -Character, Capacity and Capital

Character : It means trust of banker on borrower. Generally, people with good mental and moral character will have good credit character.

Capacity : It is related to the capacity of individual to clear loans.

Capital : It implies availability of money with the borrower.

Seven Ps' of credit : They are the principles of :

(1) Productive purpose, (3) Productivity, (5) Phased disbursement,

(2) Personality, (4) Proper utilization, (6) Payment and (7) Protection.

INSTITUTIONAL LINKAGE FOR THE SUPPLY OF CREDIT

The Reserve Bank of India

The Reserve Bank of India was established on the 1st of April 1935 and was nationalized on the 1st of January 1949 with the responsibility of providing its resources for agriculture through its agricultural credit department and the rural planning credit cell of the RBI which were taken over by NABARD consequently. The main activities of the RBI are :

1. To provide adequate assistance to the co-operatives for ensuring the timely and adequate flow of credit for agriculture.

2. Under the developmental and promotional functions, the RBI has made efforts to strengthen the co-operative structure through the reorganizations of state/central co-operative banks, rehabilitation of administratively and financially weak central co-operative banks, strengthening of the PAC's and organization of suitable training programmes for the personnel of co-operative institutions.

3. The RBI re-oriented the lending policies of land development banks and, streamlined their procedure of lending and also implemented various norms for the regulation of advances by these banks to prevent the over dues.

4. The RBI has taken steps like the introduction of crop loan system conversion of short and medium term loans for the re-orientation of the lending policies of the co-operatives etc.

5. The credit guarantee scheme was formulated by the RBI to provide inputs to commercial banks for extending their facilities to smaller borrowers on a priority basis, to be implemented through the Credit Guarantee Corporation of India.

6. For the purpose that the growth of bank credit may be linked to the needs and as an additional measure of credit regulation, the RBI introduced the Credit Authorization Scheme to regulate the advances by the scheduled banks.

Thus, the RBI plays an important role for the availability of institutional credit for the promotion of agricultural and rural activities.

The National Bank of Agriculture and Rural Development

The NABARD was established on the 12th of July 1982 with the following main objectives :

To serve as an apex refinancing agency for institutions providing investment credit and production credit in rural areas.

Co-ordination of rural financing activities.

To initiate measures towards institutional building.

To maintain links with the RBI, Government of India, State Governments policy formulation.

Monitoring and evaluation of various projects.

Promotion of research in banking, agriculture and rural development.

To improve the quality of lending through proper control.

To ensure credit to weaker sections.

To support programmes and projects of the Government of India.

NABARD has been managing its financial resources through borrowings from the Central Government or any other institution approved by the Government, from the Reserve Bank of India, by issuing and selling bonds and debentures carrying interests guaranteed by the Central Government, by accepting deposits from the Central Government, from the State Government, local authorities like SLDB's, SCB's, CB's or any individual approved by the Central Government and by borrowing foreign currency from any bank or financial institute in India or elsewhere with the prior approval of the Central Government in consultation with the Reserve Bank of India.

THE MAIN FUNCTIONS OF NABARD ARE AS FOLLOWS

1. The NABARD provides refinance, loans and advances for agriculture and rural development for the disbursement of short term, medium term and long term loans to the institutions like the State Co-operative Banks, Regional Rural Banks, State Land Development Banks, Scheduled Commercial Banks or any other institution approved by the central Government, short term credit is given for distribution of fertilizers, seeds etc., agricultural operations, marketing of crops, any other activity for the promotion of agriculture, working capital requirements for co-operative sugar factories etc. Medium term or long term credit is given for minor irrigation, conversion loans, land development, farm mechanization, dairy, poultry etc. NABARD provides loans to the State Governments for share capital contribution to co-operative credit societies under long term operations from the rural credit fund.

2. NABARD plays an important role in the formulation of policies related to credit for agriculture and rural development and has been assigned by the RBI the responsibility of coordinating the efforts of various agencies involved in rural credit planning.

3. NABARD is also involved in the healthy functioning and growth of various organizations dispensing credit like the rehabilitation of PAC's into viable units, rehabilitation of weak banks, provision of support to the RRBs ,commercial banks etc.

4. NABARD monitors and controls the activities of RRB's, SCB's CCBs etc. through regular inspections.

5. NABARD has the training institutions of its own to meet the training needs of the banks involved in lending for agriculture rural development etc. Research projects related to credit for agriculture and rural development are also being undertaken by NABARD.

Thus the NABARD has been functioning as an apex institution for the provision of agricultural credit, developing an effective backward linkage to agricultural enterprises.

COMMERCIAL BANKS

The commercial banks forms the core of the organized banking system and constitute quantitatively the most important group of financial intermediaries in the country, comprising both scheduled and non-scheduled banks, Deposits, paid up capital and borrowings from the Reserve Bank of India from the resources of the commercial banks. Commercial banks are the most important intermediaries for promoting and mobilizing the savings and for allocating investment among the different productive sectors. The short term and medium term credit needs of both industry and agriculture are met by the commercial banks and they also help finance developmental plans by investing funds in the government securities. Initially, the commercial banks were concentrating only on the financing of the trade and industry. However, with the nationalization of the banks (1969), they are now actively involved in the disbursement of agricultural credit. On account of the branch licensing policy adopted by the RBI, the rural branches of the commercial banks account for a large percentage of the total network and the Agricultural Development Branches, Gram Vikas Kendras and Rural service centers were set up to cater exclusively to the needs of agriculture and the allied activities. Under the Lead bank scheme all districts were allotted to commercial banks who were entrusted with the responsibility of preparing credit plans for their lead districts. The village adoption scheme was formulated by commercial banks to carry out lending operations in rural areas on sound and scientific lines. Thus the commercial banks are contributing significantly to the development of agriculture.

THE REGIONAL RURAL BANKS

The Regional Rural Banks were established in 1975 to provide direct loans and advances to the small farmers, rural artisans, agricultural laborers etc. to grant loans and advances to the co-operative societies, co-operative farming societies, primary agricultural credit societies or farmers service societies for the purpose of agricultural operations and other related purposes and to grant loans and advance to artisans, small entrepreneurs and persons of small means engaged in trade, commerce or industry or other productive activities within the area of operation of the concerned rural bank. The RRB's were set up in such areas where the credit availability was inadequate, dominance of weaker sections and the agricultural developmental potential was encouraging. The management of the RRB's comprise of a Chairman appointed by the Central Government, three Directors nominated by the Central Government, two Directors nominated by the State Government and not more than three Directors nominated by the sponsoring banks. There are around more than 200 Regional Rural Banks functioning with the deposits of around 2500 crores. Refinance facilities are provided by the sponsoring banks to the respective RRB's in addition to the provision of managerial assistance. The RBI also provides refinance facilities to the RRB's through NABARD. Several other concessions are also being provided by the RBI, Thus the RRB's are expected to combine the service into of co-operatives on one hand and the professionalism of national banks on the other.

CO-OPERATIVE CREDIT

The co-operative set-up for the disbursement of credit can be divided into the short term structure comprising of the PAC's the DCCB's and the SCB's and the long term disbursement comprising of the primary and central land development banks.

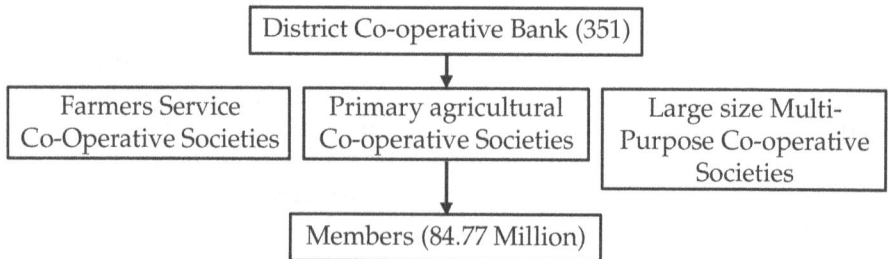

District Co-operative Bank (351)

| Farmers Service Co-Operative Societies | Primary agricultural Co-operative Societies | Large size Multi-Purpose Co-operative Societies |

Members (84.77 Million)

1. Primary Agricultural Credit Society

The primary agricultural credit society is an organization of villagers for mutual help and co-operation to meet the common economic requirements and to increase the agricultural production. The main functions of the PACs are:

(a) Supply of various agricultural inputs to the farmers like improved seeds, fertilizers, pesticides, agricultural implements etc.

(b) To provide short term credit for the purchase of farm requirements and the medium term credit for reclamation of land, sinking of wells.

(c) To provide the household requirements like sugar, kerosene and other essential items.

(d) To provide the agricultural machinery on hire.

(e) To attract the deposits of the members and to utilize it effectively.

The credit societies normally formed on the village community basis and the main sources of funds are the share capital from both the individuals as well as the State Governments and deposits of the members. Thus the PAC's form the grass roots linkage to the agriculturists for the supply of credit as well as the other requirements.

2. The District Central Co-operative Society

The primary agricultural credit societies are federated together at the district level to form a higher tier of the rural agricultural credit called the District Central Co-operative Banks. The sources of finance are the share capital contributed by the member societies and also the share capital contribution by the Government. The main functions.

(a) To provide the credit requirement to the primary agricultural credit societies.

(b) To provide for the working capital assistance to the agro-processing units.

(c) Mobilization of deposits from the public to undertake the different developmental activities.

3. The State Co-operative Bank

The district central co-operative banks are federated to form the state co-operative banks at the state level. The main functions are : To co-ordinate the various policies. Provision of credit to the district central co-operative banks. To provide for the working capital requirement of the Apex Co-operative societies. Provision of training, supervision and guidance in the area of agricultural credit.

4. The Federation of State Co-operative Banks

The federation of state co-operative banks was established in 1964 with its main objectives being : Liaison with the Government, RBI, NABARD and other national organizations on the policies affecting agricultural credit. Research, publication and consultancy on agricultural credit. Promote and project the interests of member banks in the various spheres of their activities. Provide a common forum for the member banks to deal with their problems. Thus, the National federation of Co-operative Banks functions as the apex institutional linkage to the investment credit for agriculture.

5. The Land Development Banks

The primary agricultural credit societies providing for the short and medium term finance could however not meet the requirements of long term investment finance to resource constraints and as the commercial banks were also not functioning efficiently, the land development banks were set up to meet the long term credit requirements of the farmers.

The land development banks in addition to providing the investment credit to agriculture for land development, irrigation facilities etc. are now undertaking the financing of allied agricultural activities like the dairy, poultry, piggery, sheep rearing etc. on a large scale and have also been recognized for financing under the Integrated Rural Development Programme (IRDP).

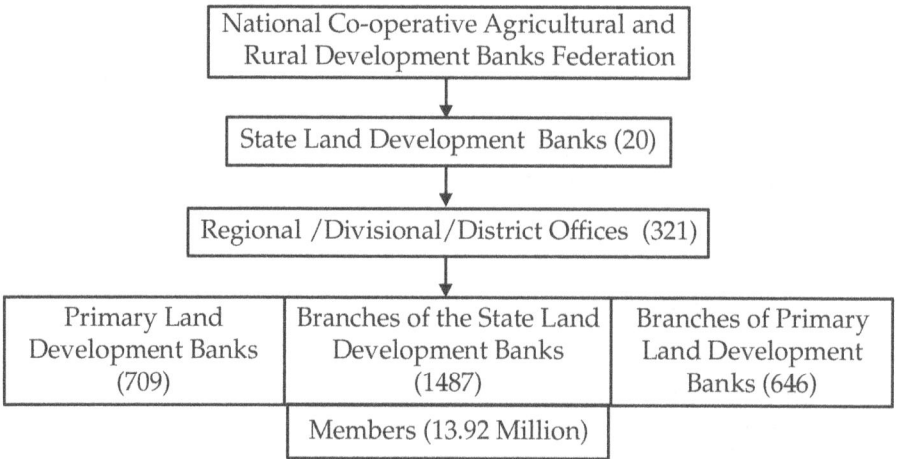

```
┌─────────────────────────────────────────┐
│ National Co-operative Agricultural and  │
│ Rural Development Banks Federation       │
└─────────────────────────────────────────┘
                    │
                    ▼
    ┌─────────────────────────────────────┐
    │ State Land Development Banks (20)    │
    └─────────────────────────────────────┘
                    │
                    ▼
  ┌───────────────────────────────────────────┐
  │ Regional /Divisional/District Offices (321)│
  └───────────────────────────────────────────┘
                    │
                    ▼
```

Primary Land Development Banks (709)	Branches of the State Land Development Banks (1487)	Branches of Primary Land Development Banks (646)
	Members (13.92 Million)	

The long tem credit co-operative structure is not uniform throughout the country and may be the central land development banks operating as departments of the state co-operative banks or the central land development banks functioning through branches as well as primary land development banks or the central land development banks advancing the loans direct to the individuals.

The state land development banks coordinate the long term credit policies, fleet debentures, give credit to PLDB's, supervise and guide the primary land development banks and liaison with the NABARD, SBI, LIC and other institutions.

The district, regional and divisional offices guide the field units for implementing the loaning policies and the procedures, carrying out the inspection of the units and verification of the credit utilization and the coordination with the other development agencies. The primary land development banks provide for the investment credit to the members. The National Co-operative and Agriculture Rural Banks Federation provides the publication, guidance, consultancy on the investment credit, liaisons with the Government, the Planning Commission, RBI, NABARD, Commercial Banks, Co-operative Banks, State Bank of India, LIC and others concerned on the matters related to long term credit.

Assessing Financial Viability

The above statements generate the basic data on the cash flows in different years. We have to apply some standard methods to evaluate the financial viability of the proposed project. There are three commonly used criteria for evaluating a project proposal for financial viability (1) Pay Back Period (PBP), (2) Net Present Value (NPV), and (3) Internal Rate of Return (IRR). These are elaborated below.

Pay Back Period (PBP) The payback period is defined as the time period within which the initial investment on the project is recovered by the unit in the form of revenues. To put it differently this is the length of time between the initial investment on the project and the time when this initial investment is completely recovered from the net yearly revenues. In symbolic terms if the net yearly cash flows are same every year, we can express the pay back period as follows:

$$P = \frac{1}{\Sigma C}$$

where, P is the pay back period

I is the initial investment and

C is the yearly net cash inflow.

If the net revenues are not uniform every year, we simply go on adding the yearly net revenues till they equal the initial cost. Once again we can write it as:

$$R_1 = I$$

Usually, the investor has a time period in mind within which he would like to recover his money and this time period is referred to as the desired or threshold pay back period. The pay back period of the project is then compared with this threshold pay back period and the project is accepted if the payback period is less than or equal to the threshold payback period.

There are basically two limitations to this investment criteria. It does not take into account all the cash flows generated over the time horizon or the project life and, second, the measure does not also take into account the time value of money.

The concept of time value of money can be illustrated with the help of an example. Suppose we take a case where the going interest rate is 10 per cent a year and a loan of Rs.1,000 is being considered, then Rs. 1,000 lent on January 1st will be repaid on the following December 31st with Rs. 100 as interest or a total amount of Rs. 1,100. Now, suppose the borrower wants to keep the money for two years; he has to pay 10 per cent for the use of the money for the first year and also another 10 per cent for the use of the money in the second year. In addition, he must pay the interest on the amount he would have paid to the lender at the end of the first year. This is the concept of compound interest and an example is given on the following page.

We can see from the illustration that Re. 1 received at the end of the first year is not the same as the money in the end of the first year because if one rupee is available today it can be invested at the rate of 10% and at the end of fifth year it will become Rs. 1.464.

Year	Amount at the beginning of the year	One plus interest rate	Amount at the end of the year
1981	1000	1.10	1100.00
1982	1100	1.10	1210.00
1983	1210	1.10	1331.00
1984	1331	1.10	1464.10
1985	1464.10	1.10	1610.50

Thus, we can say Rs. 1.464 at the end of the fifth year is equal to Re. 1 today. Symbolically, we can say that a rupee received at the end of nth year is equivalent to Rs. $1/(1+r)n$ now. These discount factors are tabulated for ease in computations. The other two investment criteria (net present value and internal rate of return) are known as discounted cash flow methods because they discount the future receipts to make them comparable to present receipts.

Net Present Value (NPV) : The net present value of investment is calculated by taking a discounted sum of the stream of net income during the expected life of the project. For the reasons explained in the preceding paragraph it is necessary to discount the future stream of net income because costs and returns in different time periods are nt strictly comparable. In symbolic terms we can express the NPV of a project generating net cash flows of R_1, R_2, R_3...., R_n for n years (the stream of net cash inflows minus the initial cash outflows) as follows:

$$NPV = \frac{R_1}{(1+r)} + \frac{R_2}{(1+r)^2} + \frac{R_n}{(1+r)^n} - I$$

where, *r* is the per cent discount rate. The investment is considered sound if the NPV is positive. A negative NPV indicates that the project is not worth considering at a given discount rate.

Internal Rate of Return (IRR) : The internal rate of return is that rate of return which makes the net present value equal to zero. Thus, it is the discount rate which makes the initial investment in the project equal to the discounted net returns from the investment during the entire life of the project. In symbolic terms we can express it as:

$$\frac{R_1}{(1+r)^2} - I = O$$

where, R_1 are the net returns in each of the n years and $(1 + r)_1$ is the discount factor. In this case R is to be estimated such that the discounted net returns from the project equal zero. The estimated value of R is then compared with the cost of capital. If the internal rate of return (IRR) is higher or equal to the minimum desired yield or rate of return, then we accept the project proposal. Otherwise, we reject the project and as a corollary to this between the two projects if one yields higher IRR than the other, the first one is preferred to the second.

SOME OTHER FINANCIAL RATIOS

To enable the entrepreneurs to form a judgment about the efficiency of the enterprise, creditworthiness, and his return on key aggregates, some financial ratios can be computed from the projected financial statements of the unit.

Efficiency Ratios : As the name indicates these ratios provide information on the efficiency of the proposed agro-industrial unit. Some of the important ratios in this regard include inventory turnover and operating ratios.

Inventory Turnover Ratio : The inventory turnover is computed as follows: sales are divided by the inventory. This ratio measures the number of times the unit turns over its stock each year and indicates the stock of inventories required to support a given level of sales. It is generally observed that in agro-processing industries this ratio is lower compared to other industries because of their seasonal operations. It is also observed that agro-industrial units have to generally keep several months of inventories to support their processing operations and production schedules. A lower ratio also implies that sizeable amount of funds are locked up in inventory for long times. It is to be judged whether the computed ratio is reasonable, given the conditions and practices prevailing generally in that industry.

The Operating Ratio : This ratio is computed by dividing the operating expenses with the revenue. This ratio indicates the ability of the unit to control operating costs including overhead expenses. This ratio is generally used in comparing the performance over the time of the same unit or comparing the performance of the proposed unit with that of other units. If it is observed from the projected statements that the ratio increased over time, it implies either the cost of raw materials is increasing or the labour cost is increasing or there are wastages in the production process and/or there is substantial competition necessitating a reduction in prices. When the operating ratio becomes very high the unit may have difficulty in making an adequate return. On the other hand, if this ratio is very low, one has to examine whether some of the cost items have been omitted.

Income ratios : An entrepreneur has to examine whether the projected enterprise would be able to provide or generate resources for reinvestment and growth and its ability to provide a reasonable return on investment. The most important income ratios are : (1) return on sales, (2) return on equity, and (3) return on assets. These income ratios vary from year to year. Therefore, these need to be computed for each year from the projected statements for the unit.

Return on Sales : This ratio indicates the operating margin of the unit on its sales. If the return on sales or the operating margin is low, it implies that the unit will have a large volume of sales in order to earn an adequate return on investment. This ratio is usually used for comparison purposes with the units in the same industry or the same unit over time. Since the acceptable level of the ratio varies from industry to industry it is meaningless to compare this ratio across industries.

Return of Equity (ROE) : This ratio is computed by dividing the net income after tax by equity. This ratio helps the entrepreneurs to compare various investment opportunities and select the project proposal which yields a satisfactory return on investment.

The Return on Assets : This ratio indicates the earning power of the assets of the proposed unit and it is computed by dividing the operating income by the value of assets. If the proposal is to be acceptable for entrepreneurs, the return on assets should exceed the cost of capital; otherwise the project is not viable from the point of view of the entrepreneurs in the sense this ratio is close to the calculations or indicators of the financial viability in the previous section.

The Creditworthiness Ratios : These ratios indicate the degree of financial risk inherent in the enterprises before going for them. These ratios also indicate the type of financing and terms the proposed unit would require so that it may be able to survive even the adverse circumstances. Some of the important ratios in this category are the current ratio, the debt-equity ratio, and the debt-service coverage ratio.

5

AGRI-BUSINESS MANAGEMENT

NATURE, SCOPE AND IMPORTANCE OF AGRI-BUSINESS MANAGEMENT

Concepts of Agri-Business

Agri-Business has reference to the production, processing and marketing of the farm products. It covers the farm suppliers, farmers, traders, processors and retailers. It deals with the institutional arrangements, which affects the various stages of Agri-Business System. The Govt. agencies, co-operative, contractors, financial partnerships, vertical and horizontal integration and backward and forward linkages are the areas which are included in the study of agri-business.

MANAGEMENT

Management is the art and science which deals with organizing and operating a series of activities which culminates in a coordinated productive whole.

The study of agri-business management is interdisciplinary. It makes use of the concepts and techniques from the disciplines of management, industrial economics, agricultural economics and development economics. The micro and macro dimensions of agri-business management and development in a system framework are studied in these disciplines.

HISTORY OF AGRI-BUSINESS MANAGEMENT

In modern times, agri-business as a concept was born in Harvard university in 1955 with the publication at a book "A concepts of agri-business" under the joint authorship of John Davis and R. Goldberg. In his book "Introduction to Agri-business management". J.D. Drilon has given details.

Definitions of Agri-Business given by different authors

John H. Davis

Agri-business has been defined as all the activities concerned with agriculture including farming, management, financing, processing, marketing, growing of seeds and nursery stock, manufacture of fertilizers, chemicals, implements, processing machinery and transportation equipment and the process of transportation itself.

E. Paul Roy

The science of agri-business was defined by E. Paul Roy as the coordinating science of supplying agricultural production inputs and subsequently producing, processing and distributing food and fiber.

Agri-business comprises of all farm inputs, farm production and food and fibre processing and distribution entities involved in the production and distribution of food and fibre products to domestic and international consumers.

Prof. R. Goldberg

According to him an agri-business commodity system encompasses all those involved in the production, processing and marketing. Such a system includes farm suppliers, farmers, storage operator, processors, wholesalers and retailers involved in commodity flow from initial inputs to the final consumer.

Thus, the concept of Agri-business and Agri-business management encompasses the entire agricultural sector including fishery and forestry. That part of industrial sector, which contains the sources of farm supplies or the processors of farm products, is also included in agri-business.

Thus, agri-business is wider concept. It includes production (where farm management is needed) processing (which is one part of agro-industry) and marketing. Thus the term agribusiness includes practically all the facets of agriculture and the business related to it.

NEED TO STUDY THE SUBJECT OF AGRI-BUSINESS MANAGEMENT

There is no doubt that the agri-business activities are on the increase. The liberalization policies of the Government and the establishment of WTO have created more opportunities for globalising our agriculture.

There are clear indications that certain sectors such as floriculture, aquaculture, poultry, processing of fruits and vegetables are reaping the benefits of advanced technology. The entrepreneurs or organizations engaged in such ventures are on the look for competent and trained agri-business managers.

Also, commercialization of agriculture calls for specialized production, post-harvest management, and expansion of processing, transportation and packaging activities and marketing of products both in domestic and international markets. Hence, there is a need to study the subject of Agri-Business management as it has become important in different enterprises and organizations.

SCOPE OF AGRI-BUSINESS

Agri-business as an industry has got tremendous scope as it covers not only wide range of activities but also multifarious magnitude of activities. The scope of the agro-business industry can be classified into four main categories—

1. The Resource supply sector
2. The product marketing sector
3. The processing sector, and
4. The whole sale and Retail sector

1. **Resource supply sector:** Several industries can be set up in the resource supply sector such as the feed manufacturing sector, manufacture of the farm machinery, industries dealing with the seed production and supply, industries supplying energy, organizations providing the requirements, industries supplying fertilizers, agril. chemicals and the other agricultural services.

2. **Product marketing sector :** There exists tremendous scope in the marketing of Agril-Commodities. The commodity marketing organizations have a lot of opportunities in the agri-business sector. Several organizations provide facilities for the storage and warehousing of the Agril- Commodities including the specialized storage such as the cold storage etc. Transportation is most important service sector for agril-commodities which takes the product from one place to another. Thus, the product marketing sector provides or having vast opportunities to facilitate the agricultural production.

3. **Processing sector :** There exists vast scope in the processing sector. Several food processing industries such as the sugar, dairy, poultry and several other industries based on the processing of horticultural and floricultural commodities form the mainstay of the Indian economy. The oilseed processing industry for the extraction of oil and fats also play a very important role in agribusiness. The alcoholic beverage industries, textile industries, wood and paper industries, tobacco industry and such other several agro-based industries are playing major role in the Indian economy and thus agribusiness in these industries has got very wide scope.

4. **Wholesale and Retail Sector :** The whole sale and the retail sector is also dependent on several agro based industries such as the food, soap, paint, leather, beverages, textile, wood and paper, tobacco, chemicals etc. This sector provides a wide range of services and thus vast scope.

IMPORTANCE OF AGRI-BUSINESS MANAGEMENT

The ongoing structural changes in the economy have resulted in major shifts in the Indian agricultural scenario. The commercialization of agriculture opening up vast opportunities for value addition and also packaging and exports of agricultural products with high levels of technology. The policies of globalization have taken Indian agriculture in to the global village, opening of opportunities as well as great challenges. These and other forces of change are placing significant demands for managerial skills in the Agri-Business sector.

With little focus on development of managerial skills in the undergraduate curriculum, our young agricultural engineers find it difficult to function effectively as managers in Agri-Business firms, which have emerged as their most important employers. The study of Agri-Business management is response to this situation and it aims at enable meritorious agricultural engineers to acquire the critical competencies to function as effective Agri-business managers.

CHARACTERISTICS OF FARMING AS A BUSINESS

Farm means a piece of land where crops and livestock enterprises are taken up under a common management and has specific boundaries. The term farming refers to the methods and practices followed in production of crop and livestock.

For successful organization of a business and for optimum use of resources, it is important first to understand the characteristics of that industry or business as the characteristics of farming influences the actions of farmers as a businessman. The special characteristics of farming that influence the action of farmer are as follows.

CHARACTERISTICS OF FARMING AS A BUSINESS

Farming as a business is having many distinguishing features from most of the other industries in their management, methods and practices. The major differences between farming and most other industries are:

1. Primary forces of production
2. Size of the production unit
3. Dependence on climatic factors
4. Frequency and speed of decisions
5. Changes in prices
6. Standardization of practices
7. Turn over
8. Financing
9. Inseparable home and farm business
10. Nature of variable and fixed costs

11. Demand for farm products

12. Time rigidity in consumption of Agril. Products.

1. PRIMARY FORCES OF PRODUCTION

Agricultural production is primarily biological in nature. Farming deals with living enterprises, i.e., plants and animals. A slight change in the environment may cause serious difficulties. Unforeseen changes in environment such as plant or animal diseases and storms can cause a considerable damage. Most of the other industries are less affected by such circumstances. Farmers have little control over production due to agriculture being seasonal and biological in character. Management practices in farming therefore be adjusted to meet these conditions.

Agriculture requires a far larger proportion of land in relation to other factors of production as compared to other industries. The chief differences between agriculture and industries are as - diminishing returns at an earlier stage, wide scatter of production, greater importance of systems of land tenure.

2. SIZE OF PRODUCTION UNIT

Farming is, comparatively a small size business and thus provides a little scope for division of labour. A farmer is generally a labourer and a capitalist since he works on the farm, on which all his accumulated wealth and saving get invested. Most of the other industries, however, are organized on a large scale. Because of this difference in size of unit, management problems of agriculture vary greatly from those of many other industries.

Farming is not basically suited to large scale operations because it is concerned with living things, plants and animals scattered over space which need prompt attention and personal care of the farmer.

3. HEAVY DEPENDENCE ON CLIMATIC FACTORS

Weather is a very important but unpredictable variable in all farming operations.

Any sudden change in the weather, i.e., temperature, rainfall or humidity would involve readjustment of entire day's work, perhaps the entire week's, work on the farm. However, the weather and the changes in it may cause less or minor inconvenience in other industries. Thus, farming is subjected to great risk and uncertainly on account of excess or failure of rainfall, draught, hailstorm, floods, insect, pest and diseases. As a result of dependence on climatic factors, the management practices on farming must be more flexible than in other industries.

4. FREQUENCY AND SPEED OF DECISIONS

Farming requires many and speedy decisions on the part of a farmer and farmer workers. When a sudden rain floods all the croplands, there is no time to discuss of paying the double wages. The entire field must be drained immediately to save the growing crops even by paying more than double wages to the labourers.

5. CHANGES IN PRICE

Agriculture prices and production usually move in opposite directions. Because of the effect of climatic and biological factors, a relatively large volume of production of a given farm commodity is usually followed by a decline in prices; and a smaller volume results in increase in prices. This difference is due to fact that in short run agriculture has little control over the volume of production resulting in inelasticity of supplies. Demand for agricultural commodity is also relatively more inelastic. Even in the long run it is comparatively more difficult to adjust the volume of production to price variations. As against to this industries checks over production by adjusting manufacturing schedule.

6. STANDARDIZATION OF PRACTICES AND PRODUCTS

Standardization of practices and products is impossible in agriculture. However, the non agril-industries are characterized by production of large volume of highly standardized single branded products. Because of use of machines and trained personnel large volume of produce can be produced exactly the same in size, form and quality. Such standardization is impossible in agriculture. The management problems of farmers thus different from those of other industries because marketing difficulties involved in small volume of indifferent qualities products.

7. TURNOVER

Rate of capital turnover is the gross returns on a farm as a percentage of total capital investment.

$$\text{Rate of capital Turnover} = \frac{\text{Gross income} \times 100}{\text{Total farm Assets}}$$

The rate of capital turnover is basic measure of efficiency in the use of capital employed by the business. A higher turnover rate usually means efficient use of capital. In general retail business or merchandizing the turnover is normally quick, i.e., in one to three months and total capital investment gets recaptured. In agriculture, this period varies from seven to ten years in general. The turnover rate in most other types of businesses is ten to twenty times as rapid as in agriculture. The turnover rate in agriculture is relatively slow because the production process ordinarily requires period of several years.

8. FINANCING

Agricultural financing faces different problems than those faced in financing other types of business. Investment in agriculture is comparatively more risky and expensive because farming is subject to many risks from storms, dry periods, insects, pests, diseases and such other external factors. Also, because of slower rate of turn-over, repayment provisions have to be different from those of other types of investment. The capital used in agriculture is generally periodic and spreads over a longer period credit plans, therefore, has to provide for repayment time and rates according to recovery.

9. INSEPARABLE HOME AND FARM BUSINESS

Farming is a unique industry in that it combines mode of life with the business enterprise. In any other industries, this type of close interdependence does not exist. In agriculture some farm inputs may be used partly in the home e.g. food grains, feed grains, telephone, transport vehicle etc. Agriculture is therefore often regarded as a way of life as well as means of livelihood such that socio-psychological and financial credit considerations influence its organization more than they do so in other forms of industries and business.

10. NATURE OF VARIABLE AND FIXED COSTS

In agriculture portion of the fixed cost is more in total cost. This portion of fixed cost tends to make more difficult for adjustment in production, which generally makes a production process rigidly. Their level and combination cannot, be changed quickly in response to changes in prices and economic environment.

11. DEMAND FOR FARM PRODUCTS

Agriculture is mainly concerned with production of food and raw material. It is to be expected that as standard of living improves and income increases, demand for agricultural products will increase less rapidly than that for industrial products i.e. demand for agricultural products (especially for food) is inelastic. This characteristic puts agriculture at disadvantages. Higher production may reduce prices so low that total returns might not increase or may even decrease.

12. TIME RIGIDITY IN CONSUMPTION OF AGRIL. PRODUCTS

Agricultural products are generally perishable. It is not easy to postpone their utilization or consumption partly as a result of this, and partly because of small-scale individual production, the market intermediaries between producers and consumers assume a place of particular importance. They come into provide place, time and form utility to the farm produce to make it available to the consumers in the form, at the time and place required by them. The share of producers in consumer's rupee therefore goes low in case of agricultural products.

FUNCTIONS

The fundamental functions of management are - (1) Planning (2) Organization (3) Staffing (4) Motivation/Directing (5) Controlling

(1) Planning : It is the most basic function of management. Planning encircles the selection of future course of action from amongst the alternatives for organization as a whole and department within it. Each manager plans and his course of action depend upon his function. In fact plans involve selection of enterprises goals. Plans involve determining articulately departmental objectives and programs. Plans also determine the ways to reach them.

Billy E. Goetz has stated planning is fundamentally choosing. A planning problem arises when an alternative course of action is discovered. Planning is more than decision making. Because it involves meeting legal and other requirements imposed by forces beyond control of management or manager. Planning is deciding in advance - What is to be done? When it is to be done? By whom it is to be done? Why it is to be done? How it is to be done? How to bridge the gap in between where we are standing and where we want to go ?

Some view planning as a process involving the forecasting of future events, problems and selecting courses of action to handle these foresee or anticipated problems and events. Another characteristic of planning is it involves taking the decision in advance of action.

(2) Organization : Once the planning is effected the people—the organization have to be organized towards achieving the organizational objectives so predetermined. Basically organizing is concerned with. (a) grouping the activities required to achieve the organization objective. (b) Delegating appropriate authority to the people to discharge their respective responsibilities and (c) Establishing structural relationship to enable coordination of the individual effort toward the accomplishment of organizational objectives.

(3) Staffing : Staffing is concerned with ensuring that the right type of personal is available to man and execute the varied activities, required to attain the planned activities as well as objectives of the organization. It therefore, includes manpower planning, selection process, providing training and development opportunities etc.

(4) Motivation / Directing : Motivating is concerned with directing the effort of the human beings towards implementation of the plans in such a way, as to make them to do desired. What is in the interest of attaining the organization objectives or goals. A manager's personal value and the philosophy as well as his leadership abilities are really at test in this field of motivating the subordinates or getting things done through them.

(5) Controlling : Adequate control is a Sine-Qua-nonensuring that the objectives in terms of the plan are fulfilled. It include evaluation to determine whether the planned objective or result have been achieved or not. In case there are not, there must be immediate indication as to where improvements are required. Adequate control thus leads to innovation and improvement in previously determined objectives. A critical thinking of the future, an in depth analysis of the pros and cons and logical acceptance of the best available means of attaining goal embraces a successful planning. The more the involvement, greater the commitment with a visionary approach. The result yielded is innovative, practical, pragmatic, positive contribution. Planning and control go hand in hand. The assessment of existing status assists to know what corrective steps are required to be undertaken for bridge the gap between where we are standing and where we want to go. The management process approach thus provides the necessary mental frame work for analysis of the job of managing by using the contributions of the various schools of thoughts of management.

PLANNING

Planning is the most basic of management function. Future is not known to any one and therefore planning is essential. When we think deeply, conceptualize pros and cons, consider tangible and intangible aspects, determine the controllable and noncontrollable factors, we are made aware of the possibilities and the probabilities in the future. This leads to think the alternative solutions. We measure the same and the best possible answer to the problem is known to us. Therefore, we can say that planning is choosing of future course of action from amongst the alternatives for (a) organization as a whole (b) department/section within it.

Every manager irrespective of the fact at which level he is in the organizational hierarchy, plans his course of action depends upon his function. Plans embrace choosing. This encircles selection of enterprises goals, selection departmental objectives and programs and determining ways to reach them.

Planning is concerned with the conscious determination of the courses of action, required to achieve predetermined objective. Planning is viewed by some, as decision making since it has to decide in advance about—What is to be done? When it is to be done? By whom it is to be done? Why it is to be done? How it is to be done?

In fact decision making is the choice in between almost right and probable wrong with reference to the situational factors. The success and effectiveness of any plan, programme, principle, job depends upon the ability to marshal the resources and support from top.

Definition : George Terry defines planning as : Planning is the selecting and relating of fact and the making and using of assumptions regarding the future, in the visualizations and formation of proposed activities believed necessary to achieve desired results.

Billy E Goets stressed that planning is fundamentally choosing.

A planning problem arises when an alternative course of action is discovered. In fact there are no problems. There are questions. And problems are created if we fail to answer those questions. Planning is more than decision making because it involves meeting legal and other requirements imposed by forces beyond the control of a manager. Planning makes it possible to occur things which otherwise do not happen. Exact future cannot be predicted. Factors beyond control may interfere with the best laid plans. Therefore, without plans events are left to chance. Planning is in fact an intellectual process. This intellectual process comprises of conscious determination of course of action, basics of decision on purpose, consideration of facts and consider estimate. In an economic, social, technological, political era planning is essential for survival.

THE NATURE OF PLANNING

The essential nature of planning can be understood through four basic planning principles.

   (i) **Contribution to objectives :** Plans alone can not make an enterprises successful. Plans focus action on purpose. They focus which action will tend to realization of an ultimate objectives; which tend away, which will likely offset others, which is irrelevant. Managerial planning seeks to achieve consistent, coordinated operation focused to desired end. If there is no planning; random activities results in fruitless realizations.

  (ii) **Primacy of planning :** Planning is primarily the requisite of organization, staffing, directing and controlling. Planning establishes objectives and therefore group activities are charted. Planning and control are inseparable.

 (iii) **Pervasiveness of planning :** Planning is a function of every manager. The character and the breadth of planning is directly proportional to authority and nature of policy and place. Not only this but your ultimate satisfaction depends on the ability of plan.

 (iv) **Efficiency of plan :** The efficiency is measured in terms of amount of contribution to objectives, cost factors and unsought consequences required to formulate and operate.

ADVANTAGES OF PLANNING

There are several benefits of planning. The same are stated here below.

   1. Planning increases the organisation's ability to adopt to future eventualities.
   2. Planning helps crystallize objectives.
   3. Planning helps a relatedness among decision.
   4. Planning helps the company to remain more competitive in its industry.

5. Planning reduces unnecessary pressures of immediate nature.

6. Planning reduces mistakes and oversights.

7. Planning ensures a more productive use of the organisation's resource.

8. Planning makes control easier.

9. Planning enables the identification of future problems.

10. Planning helps organisation to progress in the manner most suitable to its management.

11. Planning can help the organization secure a better position or standing.

12. Planning increases the effectiveness of a manager.

Limitations of planning : Though there are several advantages of planning's, yet we find there are certain limitations like. The effectiveness of the plan depends upon the correctness of assumptions, Planning is expensive, Planning delays action and Planning encourages a false sense of securing.

CHARACTERISTICS OF A GOOD PLAN

Maleolm Pennington has highlighted the following characteristics of a good plan.

(1) Involve top management only at key points in the planning process.

(2) Involve line executives in developing the plan.

(3) Do not look for the perfect answer.

(4) Planning must provide realistic schedules targets and alternative ways to achieve them.

(5) Planning should start on a small scale and be expanded only when the executive have learned the technique and have been convinced of their usefulness.

STEPS IN PLANNING

Many a time planning appear simple but it is a complex area. In order to have as much possible prediction following activities are to be contemplated.

(i) Crystallizing the opportunity or problem/Gathering facts.

(ii) Securing and analyzing necessary information/Analyzing facts.

(iii) Establishing planning premises and constraints/Forecasting change

(iv) Ascertaining alternative course of action and plans/Setting goals.

(v) Selecting the optimum plan/Setting results

(vi) Determining derivative plans/Developing alternative.

(vii) Fixing timing of introduction.

(viii)   Arranging future evaluation of effectiveness of the plan/Evaluating progress.

Types of Plan : To illustrate the breadth of planning; they are classified. There are different ways in which plans are classified. One approach is classifying according to the time dimension—Short term plan, Medium term plan and Long term plan.

Plans classified in terms of function—Production planning and Marketing planning.

PLANNING PROCESS

(1) Objectives : Objectives are nothing but aims. These are aimed at the end towards which activity is aimed. In fact it is not the ultimate end point of planning but also aimed at organization and organizing, staffing, directing, coordinating and controlled. It includes departmental goals. Let us take an example. Earning profit is the goal of an organization. So certain departmental manufacturing goal is set. This will involve lower possible cost but high quality production.

In fact establishment of objectives, is the first step in the planning process; towards which the activities of the organization are to be directed. Objectives results in channelising of activities, actions towards accomplishment of certain results described as goals, aims and purpose.

Now a days the phenomena MBO (Management By Objectives) is attracting every ones attention. It is also described as Results Management or Management by results. It is aimed at increasing the effectiveness of managers by placing the responsibility on each manager for achieving results for the parts of the organizational activities. The objectives are set by each manager, either with participation, or through consultation or through the "Top Down Approach". Whatever method is used, following are some of the essentials.

  (i)   The individual and the departmental objectives must dovetail in the overall organizational objectives.

 (ii)   Objectives must be clearly defined and communicated.

(iii)   Objectives and goals must be reasonably attainable.

(iv)   Objectives fixed for an individual must consider uncontrollable factors.

 (v)   Objectives should be reviewed periodically for necessary changes.

ADVANTAGES IN MANAGEMENT BY OBJECTIVES (MBO)

  (1)   They provide a basis for planning and for development of other plans such as policies, budgets and procedures.

(2) They result in a better appreciation of what the organization is attempting to achieve and give meaning and direction to people.

(3) Coordination can be achieved.

(4) Standardization assists in exercising control.

(5) They provide a motivation device.

(6) They give a direction and force management to think ahead.

(2) Policies : Policies are general statement or understanding which guide or channel thinking in decision making of subordinated. Policies delimit area within which decision is to be made. Policies tend to predecide issues, avoid repeated analysis, give unified structure to other types of plan. Policy formulation constitute a type of planning, where policies from the continuous framework, within which, the persons in the organization operate. Policies thus constitute the guides to action or decisions for the various persons in the organization; reflecting the organizations official attitudes, towards future behavior of employees within it.

Koontz and O'Donnel divide the source of policy into the following four types.

(a) **Organisational policy :** They normally originate from the top management itself. Basically, they flow from the organization objectives. Other derived policies may be developed at subsequent levels, depending upon the extent of decentralization.

(b) **Appealed policy :** It means decisions given in case of appeals in exceptional cases upon the management hierarchy.

(c) **Implied Policy :** Is meant, policy which eliminate from conduct.

(d) **Externally imposed Policy :** From outside the organization such as Government control

Policy hierarchy : Policies are made at different levels in the organization itself. There thus, exists hierarchy of policies, just like a hierarchy of objectives and plans. The important policies are really administered at the top and have the maximum scope. The importance of these policies, depends on the level of their operations. The policies at lower levels must naturally consisted and support the policies formulated at the higher level.

The hierarchy of policy-making and policy-implementation could be broadly as follows:

Owners/Share holders Board of Directors Chief Executive/Top Management Middle Level/Jr. Level Management Supervisors Workers

PRINCIPLES OF POLICY-MAKING

The adoption of the following principles make for sound and effective policy making:

1. The policy statement should be definite, positive, clear and understandable to everyone in the organization.

2. Policy statements should be reasonable, permanent and stable.

3. At the same time they are required to be flexible also. Modifications of policy must be communicated to everyone, expected to implement the revised policy.

4. Policies should be based on fact and sound judgment.

5. Policies should not prescribe detailed procedures.

6. Policies should reflect the objectives prescribed.

7. Policies should be communicated.

8. Policies should be subject to evaluation.

9. Policies should as far as possible, be stated in writing.

Media for Policy dissemination : Apart from implied policies, the two basic ways in which policies are disseminated through - Written Statements and Oral dissemination.

Advantages of written policy statement	Disadvantages in written policies
1. Exact interpretation and common understanding.	1. Inclined to promote rigid thinking.
2. Can be easily reviewed from time to time.	2. Always a time-lag for incorporation of changes.
3. Can be checked more readily for compliance.	3. Likely to be interpreted differently.
4. Become available in the same form to all concerned.	4. Confidential policy statements may leak out.
5. Can be communicated and taught to new employees more readily.	
6. The process forces the managers to think clearly.	

(3) Procedures : There are methods of handling future actions and activities. These are guide to action rather than thinking. They detail the exact manner in which a certain activity must be accomplished. The essence of procedure is the chronological sequence of activity.

(4) Rules : They are usually simplest type plans. Rules require specific action/definite action with reference to the situation.

(5) Programmes : Programmes can be defined as a complex of polices, procedures, rules, tasks, assignments and other elements necessary to carry a given course of action.

(6) Budgets. : These are the statements of expects results expressed in numerical terms. In fact it is a control device.

(7) Grand strategies : It is an overall general plan. Not only this; this is major portion of a plan. The process of deciding on objective, on the resources used and policies on acquisition, use and disposition of resources. The purpose of grand strategies is to determine and communicate through a system of major objectives and policies a picture of what kind of enterprises is envisaged. They do not outline how it is to achieve.

(8) Competitive strategies : These are plans in the light of competitors plans where conditions are same; goal is same; but two or more persons existing.

The manager must have complete knowledge of competitors plans. Then they weight own plan with reference to the complete knowledge of competitors plan.

THE IMPORTANCE OF PLANNING

Without planning decisions become random and need to adhoc choices. The importance of planning is reflected in following points.

(a) **To offset uncertainties and changes :** Uncertainties and changes make planning a necessity. Planning means selecting from amongst the available alternatives the best way to accomplish an objective.

(b) **To focus attention on objectives :** Planning is directed towards achieving enterprise goals

(c) **To gain economic operation :** Planning minimizes costs because of emphasis on efficient operations and consistency. If substitutes unwarranted work.

(d) **To facilitate control :** It helps to check subordinate accomplishment.

DECISION MAKING : Decision making is the main and the major task of every manager. It is the basic and fundamental activity of manager. It has to be done in connection with formulating plans, establishing objectives, laying dawn policies and so on.

Decision making can be defined as the selection based on some criteria of ones behaviour alternative; from two or more possible alternatives. To decide means by employing the theory of elimination; to come to certain vital conclusions.

DECISION MAKING STEPS

(1) Define and crystallize the problem

(2) Secure and analyse pertinent facts

(3) Develop alternative solutions

(4) Decide upon the best solution or optimum course of action

(5) Convert the decision into effective action

ADMINISTRATIVE PROBLEMS INVOLVED IN DECISION MAKING

Soundness of decision, Decision Environment, Timing of decision, The psychological factors involved, The extent of participation and The attitude towards making a decision.

STEPS IN PLANNING IN AGRIL. BUSINESS

With the recent technological developments in agriculture, farming has become more complex business and requires careful planning for successful operations. More correctly, the farm business needs to be reorganized to produce more efficient and profitable system by way of deploying efficient production and marketing operations, consciously with a intellect in respect of what and how to produce, when and where to buy and sell. The secret of economic success of agriculture therefore, lies with proper planning of the operations and their execution based on technological advancements, changes in physical and economic situations, price changes etc.

Farm management is a process of decision making by a farmer in running his business. These decisions are of two types. This can be described as planning and operational decisions. The planning decisions are concerned with the overall organisation of the farm business. They are long-term decisions though they need to be modified from time to time with changing situations. They are normal not to subject to alterations.

The major decision taken while operating the day-to-day activities of the farm with an objective of obtaining more profits are operational decisions. They are not long term decisions and they can be taken easily. Both type of decisions are very necessary for utilizing the available resources efficiently and obtaining maximum returns.

WHAT IS FARM PLANNING?

Farm planning is a process, which helps the farmers to choose, organize and carryout the different farm enterprises for obtaining maximum income and satisfaction.

Farm planning is a process of making decision regarding the organization and operating of a farm business, so that, it results in a continuous maximization of net returns of a farm business.

More specifically, farm planning is a process to allocate the scarce resources of the farm and to organize the farm production in such a way as to increase the resource use efficiency, the production and the income of the farmer.

PURPOSE OF FARM PLANNING

The main purpose of farm planning is to help the farmers for increasing their level of production and income by adopting scientific methods of arming. Also, to improve the organization and operation of the farms.

ADVANTAGES OF FARM PLANNING

1. Income improvements
2. Educational process
3. Desirable organizational changes
4. Determination of needed adjustments
5. Formulation and appraisal of development projects
6. Road to farm security and yard stick for agril. credit.

HOW FARM PLANNING HELPS

Farm planning helps the farmer to do the following things in an organized, systematic and effective way

1. It helps him to look at his situation and past experiences as a basis to decide which of the improved ideas and methods fit to his situations.
2. It helps him to take decisions in relation to the crops to be grown, the area to be brought under cultivation or the number of livestock to be reared.
3. It helps him to identify the credit needs both short and long term and its sources.
4. It helps him to identify clearly the various services and supplies, needs for improved plan.
5. It gives him idea about the yield that can reasonably expected from each enterprise.
6. It gives him the clear idea about the returns that may be obtained from each enterprise and farm business as a whole.

CHARACTERISTICS OF A GOOD TYPICAL FARM PLAN

1. It should provide for efficient use of farm resources.

2. The crop plan should have balanced combination of enterprises.

3. Avoid excessive risks.

4. Provide flexibility

5. Give considerations to efficient marketing facilities.

6. Provides use of up to date modern agril. methods and practices.

7. Provides programme of obtaining, using and repaying the credit.

8. Utilize the farmer's knowledge and experience and also take account of the farmer's likes and dislikes.

INFORMATION NEEDED FOR PLANNING

1. Resources available on the farm, statement of resource restrictions.

2. Output to be produced.

3. Expected prices of farm products and inputs.

4. Social, institutional and personal framework with which the farmer operates his farm business.

5. Technical information - input-output co-efficient.

STEPS INVOLVED IN PLANNING AND ALTERNATIVE FARM PLAN

In developing an optimum farm plan with the budgeting technique the following steps are generally followed.

Step 1. Preparation of the farm map

Prepare farm map depicting all the physical features such as soil types, topographical features, drainage, roads, water, channels, source of irrigation, buildings, etc.

Step 2. Inventory of limited resources

Prepare a complete list of the farm resources, which limit the size of the different farm enterprises such as land, labour, animals, buildings, machinery and liquid capital etc. This helps assessment of resource limitations and production capabilities of the farm.

Step 3. Specification of technical co-efficient of production

The farmer should obtain relevant information from various sources to learn some of the improved farming methods and practices and the various input output factors, which can be applied to local conditions. This helps to prefer most efficient technology.

Step 4. Specification of appropriate prices

Prices will need to be specified keeping in view the last year's average prices, future expectations, nature of changing technology etc. One simple prediction model that can be used to assume prices next year will be the same as they are this year.

Step 5. Preparation of enterprise profitability chart

One will have to evaluate the alternative opportunities and to select those opportunities, which make the best of the farmer's resources. The enterprise budgets are prepared by deducting variable cash expenses from gross returns of each enterprise. Then profitability ranking chart is prepared on the basis of net returns of all the possible enterprises.

Step 6. Examine the existing farm plan

Examine the present plan followed for its costs and returns and resource use pattern. List the credit needs of the farmer for different purpose.

(i)   Workout the variable costs such as hired labour, seed, water charges, fertilizer, insecticides etc. for each enterprise.

(ii) Work out the gross income from various enterprises

(iii) Work out the returns to fixed farm resources for each enterprise through deducting variable costs from the gross income. Then analyze the total returns to the fixed resources from the existing plan.

Step 7. Locate the weakness of the present plan

The various weaknesses in the existing plan will act as guidelines for bringing about improvements in the alternative plan.

Step 8. List out the risks to agril. production on that farm

Make a list of all such risks and bear them in mind in developing the alternative plan. To extent possible, provide for effective steps for eliminating or reducing such risks in particular to irrigation, drainage, pests and disease control.

Step 9. Prepare the alternative plans

Alternative plans can be worked out which may vary in the amount of risk involved, labour requirements and other features as well as probable net incomes.

Step 10. Analysis of alternative plans (to check profitability)

New plans are analyzed for costs and returns (returns to fixed resources) and the optimum plan which is most practicable on that farm situation is selected. Ideally we should evaluate alternative plans on various points such as probable income, amount of risk, involved, labour and capital requirements, etc.

Step 11. Implementing the plan

The farm planning does not end with the preparation and selection of the final plan for adoption. The most important phase is its proper execution. There may be certain difficulties in implementing the plan. It is not likely that one will anticipate all the problems that may arise. For this reason, a good plan will usually provide for flexibility. Flexibility makes it possible to alter the plan as new problems arise or new information becomes available.

ALTERNATIVE FARM PLAN

Here for preparing alternative farm plan the existing farm plan is examined of a farmer

Step I - Farm map is prepared

Step II - Inventory of limited resources:

A. Land holding (ha)		Irrigated	Rainfed	Total
(a)	owned	5.05	--	5.05
(b)	rented in	0.85	0.30	1.15
(c)	rented out	--	--	--
	Total	5.90	0.30	6.20

The soil was heavy loam with medium fertility and with 1.20 hectares patch of Sandy soil. Land is well levelled with one patch of higher level fields. Drainage is normal.

LAND USE CAPABILITY CLASSIFICATION

Kharif season

Land suitable for (ha)	Remarks
(i) Maize irrigated - 4.90	Remaining land was not considered
(ii) Cotton irrigated - 3.20	for these crops
(iii) Groundnut irrigated - 1.20	Only this much land had proper sand
(iv) Groundnut rainfed - 0.30	soil depth. Remaining land was not
(v) Sugarcane - 1.60	suitable due to difficult water approach
(vi) Kharif fodder land - 2.00	

Rabi season

(i) Wheat irrigated 5.40	
(ii) Gram irrigated 0.80	Rest of the land had heavy soil and
(iii) Gram rainffed 0.20	wheat was preferred there
(iv) Berseem 1.30	Remaining land was not favoured for this
(v) Other rabi fodder 2.5	crop due to difficult water approach.

These land use capabilities obtained were not exclusive but were mostly overlapping.

B. IRRIGATION CAPACITY

The main source of irrigation was a tube well. It takes normally 12.5 hours to irrigate 1 hectare using 5 H.P. Electric motor tube well. Irrigation was not considered a limiting factor on this farm situation. In general cases, irrigation capacity is determined for critical periods from different sources of irrigation in the following Peroforma.

Irrigation capacity during critical periods.

Source of Irrigation	During summer May-July(ha)	During Winter Dec-February (ha)	Remarks
(i) Well			
(ii) Tube well			
(iii) HCanal			
(iv) Others			

Permanent Farm Labour	Peak Season	Normal Season	Farm Animals	Number	Condition
(i) Male labour days	89	66	Bullocks	3	Good
(ii) Female labour days	45	45	Other animals	2	Good
			Buffaloes	2	Good

(E)   Farm buildings

    (i)   One shed for animals   : 15 × 30'

        One room for fodder and Implements   : 12' × 12'

    (ii)   One store for grains   : 21' × 12'

(F)   Farm Machinery and Equipment

Particulars Brand/size	Nos.	Year of purchase	Original cost (Rs.)	Estimated life (Yrs.)	Junk value (Rs.)	Depreciation/ year (Rs.)
1. Wheat drumy	1	1989	800	6	50	125
2. Spray pump	1	1990	350	5	50	60
3. Bullock cart	1	1990	3500	10	500	300
4. Oil engine	1	1990	10000	10	1000	900

G. Working capital (Capital available for use in the farm business)

	Season	Owned	Burrowed	Total
(i)	Kharif (Rs.)	3500	1500	5000
(ii)	Rabi (Rs.)	2500	1200	3700
	Total	**6000**	**2700**	**8700**

Part of the sale proceeds of rabi crops were available for investment in kharif season and vice-versa.

H. Management factor : The farmer is interested in maximizing net farm income over a long period of time. He is middle pass. He has experience of growing only arable crops. He has social restrictions on producing crop like tobacco. Considers growing vegetable more risky.

I. Consumption pattern : Commodities Required (in quintals per year)

	Maize desi	Cotton	Groundnut	Sugarcane	Wheat
(i) House use	7.50	1.00	--	4.50	12.21
(ii) Farm use	2.10	0.40	2.50	2.50	5.21

The family consumption pattern was worked out to check the provision of the minimum quantities of these commodities in the alternative plans for purpose of home or farm requirement. The farmer was interested in growing the food articles for his consumption on his own farm even if it was not profitable to raise those products on the farm.

J. FIXED ACTIVITIES

For the maintenance of dairy and draft animals, the general pattern of land needed for kharif and rabi fodder was ascertained. These acreage allocations for fodders were considered as fixed activity in each case. The acreage under fodder could. However, be reduced by producing fodders with improved techniques.

Season	Acreage	Quantity Required
(i) Kharif fodders	0.90	400 quintals
(ii) Rabi fodders	0.94	600 quintals

All these resource levels (restrictions) formed the framework within which the existing and alternative cropping plans examined.

Step III : Appraisal of the existing farm plan full information on how each resource was being utilized and what were the incomes obtained from various enterprises adopted on the farm as follows Table.

Step IV : Weakness of the Existing plan :A careful analysis of the resource use in the existing plan showed the imbalance and low level of resource use.

Existing and Alternative Plan and their Economic Analysis

Sr.No	Crop Enterprise Area in ha	Existing plan Area in ha	Returns to fixed resources Perha (Rs.)	Total (Rs.)	Alternative plan Area in ha	Returns to fixed resources Perha (Rs.)	Total (Rs.)
	Kharif season						
1.	Local Maize	1.60	645	1032	0.40	775	310
2.	Hybrid maize	0.20	1118	223.60	1.60	1377	2203.00
3.	Hybrid cotton	0.88	1200	1056	1.60	1393	2228.80
4.	Local cotton	0.18	750	135	0.10	1084	108.40
5.	Gr. Nut IR	0.28	1100	308	0.80	1613	1290.40
6.	Gr. Nut rain	0.50	800	224	0.30	1302.00	390.60
7.	Sugarcane	0.50	3085	1542.50	0.60	4776.00	2865.60
8.	Kh. Fodder	0.92			0.80		
	Sub total	4.04	--	4521.10	6.20	--	9397.00
9	Fallow land	1.36	--	--	--	--	--
	Rabi season						
1	Wheat after fallow on irri.	1.20	1313.00	1575.60	-	-	-
2	Wheat after kh. Crop irri. except cotton	2.00	1175	2350	3.20	1200	3840
3.	Wheat after Hybrid cotton	0.88	1125	990	1.80	2448	4406.40
4.	Gram irrigated	0.20	875	175	0.50	1029	514.50
5	Gram rainfed	0.20	675	135	-	-	-
6	Barley	0.20	613	122.60	-	-	-
7	Rabi fodder	0.92			070	--	8760
	Sub total	5.60	-	5348.20	-	-	8760.90
	Fallow land	060	-	-	-	-	-
	Total Returns to fixed factors			9869.30			18157.90

Improvements in the alternative plan could be brought about in the light of the following weaknesses.

1. Some crops yielding relatively lower returns per hectare, such as desi-maize in kharif and gram in rabi, covered comparatively larger areas than crops promising higher returns per hectare such as Hybrid cotton, groundnut, Sugarcane and Wheat. This was mostly due to the low input requirements of these crops but this choice was inconsistent with the profit maximization objectives.

2. Enterprise budgets showed that the practice of keeping the land fallow in kharif season was uneconomical.

3. The existing per hectare yield levels of almost all crops were much below the potential levels and could be raised through the application of improved agricultural techniques such as intensive use of chemical fertilizers water and pesticides.

4. The restricting resources such as farm family labour and operating capital remained under utilized during various months, which could be better utilized.

5. In spite of relatively higher profitability and low input requirements groundnut hecterage was much less than the maximum permitted by the land suitability.

Net gain over existing plan - Rs. 8288.60.

Step V : List of Risks to Agricultural Production on the case farm:

On this farm there was very low levels of risks and uncertainties involved. Risks of damage by insect pests and plant diseases were there, which could be suitability reduced taking up prevent spraying schedules and other measures in the alternative plan.

Step VI : Within the framework of resource restriction and keeping in view the weaknesses of the existing plan and possibilities of incorporating modern technology the following alternative plan was developed. These plans reduced the acreage under fallow land and more rational use of the limited resource was planned in respect of land labour and capital.

Step VII : Analysis of the Alternative plan (testing) here is a net gain of Rs. 8288 from the alternative farm plan. It is therefore, advocated to the farmer to adopt the alternative plan.

ORGANISATION

To attain the ultimate objectives/aims the efforts are required to be channallised appropriately. Therefore, subsequent upon the determination of the general and specific objectives and plans the succeeding step in the management process is to organize the activities of the enterprises for the fulfillment of the assignments, aims, objectives. The total task is required to be segmented into different activities may be assigned to departments. These departments are then headed by the managers to whom authority is delegated commensurate with their responsibilities to enable them to contribute to the accomplishment of the pre-selected goals. Proper organizing would assist the most effective use of both the physical assets of the business such as plant, tools, equipments, materials, supplies as well as human resources of the enterprises so that the concern may be perpetuated and the objectives those including profit be achieved.

The word organization is used basically in two contexts. Those are as below:

(i) Referring to a particular company or group of persons working together.

(ii) Referring to the organisation as a structure or as a network of specific relationship among individuals.

The word "organizing" refer to a process or a managerial function. Organising is required to generate effective group action towards predetermined objectives and goals. By proper organizing a manager expects synergism, that is, the production of a total effect for greater than the sum of the individual contributions.

ORGANISING DEFINED

Koont and O. Donnell define : Organizing as involving the establishment of authority relationships with provision for co-ordination between them; both vertically and horizontally in the enterprise structure.

Louis A Allen defined it : As "The process of identifying and grouping the work to be performed, defining and delegating responsibility and authority establishing relationship for the purpose of enabling people to work most effectively together in accomplishing objectives." Thus organization involves four basic elements namely - Identification and grouping of the work, Defining the responsibility and Delegation of appropriate authority.

Establishment of structural relationship, so that the individual efforts are coordinated to accomplish or achieve the enterprise objectives.

Oliver Sheldon defined : As the organization is a process of combining the work which individuals and groups have to perform and provide best channel for efficient, systematic, positive and coordinated approach.

Hainman has defined : As organisation is a process of defining and grouping the activities of the enterprises and establishing authority relationship among them.

Mooney and Reiley has defined : As organisation is a process of formalized and international structure of roles.

From the study of these definition it is clear that organizing is essential aspect-

(a) In order to accomplish goals.

(b) To carry out plans.

(c) Make it possible to people to work effectively.

(d) An international structure of roles must be designed and maintained.

(e) Activities must be grouped logically.

(f) Authorities must be defined so that conflicts do not arise.

However the behavioural scientist look to organization from different angle. According to their view point organization is simply human relationship in group activities. It is a social structure. It is in fact a formal and informal relationship inside as well as outside an enterprise. Essentially it is he creation and maintenance of an intentional structure of roles. Now let us consider what it meant by a social organization. The social organization is a continuous system of differentiated and coordinated human activities utlising, transforming, welding a specific se o human, material, capital, natural resources into a unique problem solving process.

The prerequisites for the success of a sound organization are generally viewed from two angles. The first angle is to angle is to view the organization as a structure. It is a network of specific relationship. (For example the superior-subordinate relationship) The second angle is to look organization as a process of identifying, classifying, grouping, assigning various activities to achieve the objectives.

Bernard has stated that human beings are forced to cooperate to achieve personal goals. There is a noticeable significance of cooperation. It can be more productive but less costly.

The organization can be a formal one and an informal one. When the activities of two or more persons are consciously coordinated towards a given goal, the concerned are able to communicate with each other, they are willing to act and share a purpose. These should be a room for discretion. This is needed for taking advantages of creative talent. It is also required for the recognition of individual likes, dislikes, capacities. Its emphasis is on understanding basic realties of any group.

STEPS IN ORGANISING

Following are the steps involved in organising namely —

1. Identifying the activities involved to attain the objectives of the enterprise;

2. Grouping of similar activities on the principle of function of departmentation with further sub-division into section and, ultimately, jobs;

3. Definition of responsibility and accountability, so that each person knows what is expected of him in terms of attainment of the enterprise objectives;

4. Delegation of the requisite authority to enable each such person to carry out his responsibility;

5. Provision of adequate physical facilities to discharge adequately the responsibilities; and

6. Establishment of clear structural relationships among individuals and groups.

IMPORTANCE OF ORGANISATION

(a) It encourages specialization and increased productivity.

(b) It is likely to result in increased productivity through avoidance of duplication.

(c) It fosters co-operation.

(d) It aids expansion and growth of the enterprise.

Types and Functions of Organisation-Responsibility, Authority and Accountability

In case you want to start a business, what possibilities are open to you? One can start a business alone or may enter into a partnership with a relative or some trusted friends of you may become a shareholder in limited company. Besides these, there are some other forms of business too. One can adapt any one of them. The chief forms of business organization are of five types. viz; 1. Individual or single proprietorship, 2. Partnership, 3. Joint stock company, 4. Co-operative enterprise and 5. State enterprise.

1. Individual or single proprietorship

The common form of business organization in India is one-man business. In agriculture and retail business, this form is general rule.

This refers to that form of business organization in which there is only one single owner of the business. He bears the entire risk of the business. If the firm earns the profit, the entire profit goes to him. In case the firm suffers the loss, the entire burden is hold by him.

The main features of this system are as follows:

(i) There is only one single owner of the business firm. He is also the manager.

(ii)   He provides nearly all the factors of production. Even the land is his own or in the alternative he may take the land on rent.

(iii) He also contributes a major portion of capital. He bears entire risk of the business and his liability for payment of debt is unlimited.

(iv)   The productive process is extremely simple and involves the use of very elementary type of machinery. The production is carried on a small scale.

Small businesses the entire world over are run under this system, e.g., agriculture, retail trade, small workshop etc. are run under the system of single proprietorship. Some of these business shops may continue to exist for decades, while most of them may close down after a short while.

2. Partnership

Partnership is that form of organization in which two or more but less than twenty partners jointly own an enterprise and agree to share the profits in a pre-determined proportion. It is not necessary for every partner to actively participate in the operation of the firm. Sometimes a partner contributes his capital but doesn't participate in the management. Such a partnership is known as " Sleeping of Dormant" partner. Partnership is based on the partnership deed. The number of partners, their membership, their capital contributions, their rights and duties and the method of sharing the profit are clearly mentioned in the partnership Agreement.

Main features of partnership business are

(i)   Every partner contributes his own capital and profit is distributed among the partners according to their respective capital contribution.

(ii)   The liability of the partner is unlimited. This implies that every partner is liable to pay the debits of the firm to an unlimited extent both individually as well as collectively.

(iii) A partner can choose to have limited liability if he desires so. His liability in that case shall be limited by the extent of his share capital.

3. Joint-stock company

It is known as the corporate form of business (or corporation). A joint stock company is an association of individuals as shareholders who are authorized by the government to run a particular business.

The joint-stock company system came into vogue in Western countries during the 19th century to take up large-scale enterprises necessitating huge capital investment. This system has become today a popular type of business organization. Large-scale commercial and industrial enterprises are generally run under the joint-stock company system.

The various features of this system are

(i) The capital of the firm is contributed by a large number of shareholders who are the real owners of business enterprise.

(ii) The liability of the shareholder is limited up to the value of shares held by him.

(iii) The policy making job of firm is entrusted to the 'Board of Directors' who are elected from amongst the shareholders.

(iv) The paid managers who worked under direct supervision of Managing Director carry out the actual management of the company.

(v) The joint-stock company has a separate and distinct legal personality from those of the shareholders.

4. Co-operative enterprise

For the elimination of the exploitation of the consumers and the workers from single proprietorship system and partnership, a new type of business organization was brought into existence. This is a co-operative form of business organization.

In ordinary language, the term co-operation means to work together but in economics it carries slightly different and wider meaning.

Co-operation may be defined as that type of economic organization in which the number of persons with common economic objectives organizes themselves on the basis of equality for fulfillment of those objectives in practice.

The fundamental objectives of co-operative enterprise are as

(i) To work together on a organized basis

(ii) This organization is purely voluntary

(iii) This organization is intended to achieve certain common economic objectives.

(iv) Nearness of members to each other

(v) Equality

(vi) Self dependence

(vii) Economy

(viii) Democratic organization

(ix) Each for all and all for each

There are several varieties of co-operatives enterprises. But the main varieties considered are as

(i)   Producers co-operative organization

(ii) Consumer's co-operative organization

(iii) Credit co-operatives

There are other co-operative societies dealing with marketing, cattle rearing, irrigation etc. Infact, the principle of co-operation be applied to any field of economic activity to protect the interest of the weaker sections of society.

The co-operative type of business organization has become quite popular now. It's superiority lies in the fact that it avoids disadvantages of both capitalism and socialism. It eliminates the exploitation of the weak by the strong as is found in capitalism. In fact co-operative organization is a compromise between two extreme economic systems of capitalism and socialism.

5. State enterprise

The state enterprises refer to these business enterprises whose ownership and management are directly in the hands of government. Such enterprises are the exclusive property of the state and are directly managed by it. For example, Railways, post and telegram department.

The number of central and state government enterprises has been constantly on the increase in India during past few years. Iron, steel, coal, fertilizers, ship building, shipping, aircraft manufacturing, heavy chemical appliances etc. are some examples of public sectors in India.

SELECTION OF BUSINESS ENTERPRISE

There are many alternative business enterprises available to a entrepreneurs, consistent with technical feasibility of such enterprises. Certain enterprises are yielding higher production returns as compared to others, because of favorable conditions. These enterprises need to be identified. For selection of business enterprise, the law of comparative advantage is helpful and for this, information of various aspects at different enterprises is essential. The relative profitability of the enterprises based on cost-returns analysis is determining the selection of business enterprises. The economic optimality also influences the selection of business.

Selection of business enterprise depends upon the locality, type of business, markets, ability of businessman and amount of capital available. Hence these factors need to be considered for a selection of business enterprise and select the business enterprise, which produce the greatest net returns.

The knowledge of six basic principles involved in making rational agri-business management provides a set of tools for selection of business enterprise. These principles are as under:

1. Principle of variable proportions of laws of returns
2. Cost principle
3. Principle of substitution between inputs of least cost combination
4. Principle of equimarginal returns or opportunity cost principle
5. Principle of substitution between products
6. Principle underlying decisions involving time and uncertainly.

ADVANTAGES AND DISADVANTAGES OF INDIVIDUAL PROPRIETORSHIP AND PARTNERSHIP

I. Individual Proprietorship

(a) Advantages

1. **Low cost of production :** Since the business is exclusively his own and no one shares the profit with him, the individual proprietor works day and night for the success of his enterprise. He allows no wastage of materials in his establishment and keeps a strict vigil on the activities of his workers. All of these results into low cost of production.

2. **Close contact with customers :** Under this system, production is carried on a small scale and as such the entrepreneur pays due attention to the individual tastes of his customers. He maintains close liaison with his customers and attempts to provide maximum satisfaction to them.

3. **Close contact with workers :** Since the entrepreneur comes into daily contact with his workers in his establishment, there is little possibility of any misunderstanding arising among them with the results that labour management relations are quite happy cordial.

4. **Promptness in decisions :** Since the entrepreneur is the sole proprietor of the business, he does not have to hold any consultation with anyone on important issues facing him. He, therefore, takes decisions promptly on the spur of the moment.

5. **Security of trade secrets :** Since the production is on a small scale and the entrepreneur himself is the owner, he can safeguard his trade secrets more effectively and successfully than would be possible under any other type of business organization.

(b) Disadvantages

1. **Unlimited liability :** Since the entrepreneur alone has to bear the entire risk of his business, he is always haunted by the fear of losses and as such is rendered unable to undertake bold experiments. He can tread only the beaten path.

2. **Limited economic resources :** In view of his limited economic resources, the single entrepreneur does not have the capacity to buy the latest modern machinery for his enterprise. He also cannot afford to spend on research and experimentation. As a result, his production costs are generally on the high side.

3. **Limits of organization and control :** The single entrepreneur howsoever efficient and experienced he may be, can run the business on a limited scale only in view of the organisational limitations.

4. **Inability to face competition from bigger units :** The single small producer cannot compete successfully with bigger units in view of his limited capital organisational resources.

Despite all these defects and drawbacks, the single proprietorship has not disappeared altogether from the business scene, because certain types of enterprises are particularly suitable for it. Agriculture, retail trading, repairing workshops, hotel-keeping etc. are particularly suitable for this type of organization.

II. PARTNERSHIP

(a) Advantages

1. **No difficulty in setting up partnership :** It is quite easy to set up a partnership firm. No permission is necessary from the authorities. The partners have simply to draw up the partnership deed and get it registered with the Registrar.

2. **Availability of larger capital :** It is possible to mobilize a large amount of capital under partnership because there are a number of partners in the firm who make capital contribution to the fund of the firm. This makes it possible for the firm to go in for large-scale production, which brings several advantages in its wake.

3. **Use of diverse talents :** Each partner may have a special aptitude or ability to perform a certain task. As such, various partners can take up work in accordance with their respective abilities and aptitudes. This will bring to the firm all the advantages of specialization and division of labour.

4. **Promptness in decision making** : Since the various partners remain in constant touch with each other, they can take quick decisions on important problems confronting the firm.

5. **Risk sharing** : Under partnership, all the partners share the risk. If per chance the firm fails, all the partners will share the losses.

6. **Check on rashness** : Since the liability of the partners is unlimited, they cannot take decisions in an indiscriminate manner. They attempt to avoid, as far as possible, unnecessary risks in the interest of solvency.

7. **Availability of easier and larger credit** : The principle of unlimited liability enables the partnership firm to borrow larger sums from the banks without much difficulty. All the partners are jointly responsible for debits of the firm.

8. **Cordial relations with workers and customers** : Under this system, the proprietors (all the partners) maintain close contact with the customers as well as the workers. This removes the possibility of any misunderstanding arising between them. Such a close contact proves conducive to the maintenance of happy and cordial relations with the customers and the workers.

9. **Efficient and economical working** : This system is conducive to the efficient and economical working of the firm. All the partners are equally interested in maximizing the profits of the firm. So they take personal interest in the successful working of the firm.

(b) *Disadvantages*

1. **Unlimited liability** : This can also be a disadvantage of partnership. The principle of unlimited liability may force the partners to avoid all sorts of risks and take to routine type of business.

2. **Availability of capital in limited quantities** : As compared to a joint-stock company, a partnership firm can raise at best only limited quantities of capital from the partners. Consequently, production cannot be carried on, on a large scale. The advantages and economics of large-scale production are not available to partnership firm.

3. **Lack of efficiency** : The responsibility of management gets divided among the partners. As is well known, divided responsibility is no responsibility. Every partner tries to pass on the responsibility to others. Consequent the efficiency of management suffers a serious setback.

4. **Delay in decision making** : Since the consent of every partner has to be taken at every step in the running of the business, there is bound to be the possibility of delays arising in decision-making.

5. Differences among partners

6. Temporariness

7. Unsuitability for large business.

PERSONNEL MANAGEMENT

Meaning-Personnel management is a generic term which means the total function of recruitment, selection, development and utilization of employees executives as well as rank and file workers. The performance of personnel management functions needs to be planned, organized, directed and controlled in the same way as the performance of all other managerial activities. It is that part of general management which is concerned with people at work.

PERSONNEL MANAGEMENT—DEFINITION

"Personnel management is the planning, organizing, directing and controlling of people for the purpose of contributing to the organizational, individual and social gains" - *Edwin Flippo.*

"Management is the development of people and not the direction of things. Management and personnel administration are one and the same, they should never be separated. Management is personnel administration" *Lawrence A Appley.*

"Personnel relation is that phase of management which deal with the effective control and use of manpower as distinguished from other sources of power. The methods, tools and techniques designed and utilized to secure the enthusiastic participation of labour, represent the subject matter for study in personnel administration". *Dale Yoder.*

All Management is personnel management as it deals with human beings its development can best be discussed in format of human development, philosophical, psychological, spiritual and physical.

OBJECTIVES OF PERSONAL MANAGEMENT

1. To design and develop an effective organization which will respond appropriately to change.

2. To obtain and develop the human resources required by the organization and to use and motivate them effectively so that they make maximum contribution to the organization's goals.

3. To maintain good relations with in an organization.

4. To meet the organization's social and legal responsibilities. The personnel management should aim at :

   (a) Attain economically and effectively the organizational goals

(b) Serving to the highest possible degree the individual goals and

(c) Preserving and advancing the general welfare of the community

SCOPE OF PERSONNEL MANAGEMENT

The scope of personnel management is very wide. It depends upon the creative attitude of the personnel administrator to bring within the fold of his activities, as many areas of administration as directly or indirectly instrumental to achieve the objectives of personnel management. The scope of personnel management is mainly confined to the following important activities :

1. Effective utilization of human resources in the achievement of organization goals.

2. Establishment and maintenance of an adequate organizational structure and desirable working relationships among all members of the organization.

3. Securing integration of the individual and informal groups with the organization and thereby, their commitment, involvement and loyalty.

4. Recognition and satisfaction of individual needs and group goals.

5. Provision of maximum opportunities for individual development and advancement.

6. Maintenance of high morale of human organization.

7. Continuous strengthening and appreciation of human assets.

Personnel management is multi disciplined phenomena and brings under its coverage number of disciplines. Some of the more important disciplines utilized in the field of personnel management are : Philosophy, Anthropology, Ethics, Medicine, Logic, History, Mathematics, Economics, Psychology, Study of management, Sociology, Political Science.

Personnel manager must know why these disciplines are so important and how they are applied to labour problems. As regards its importance, careful study of the foregoing disciplines, discloses the fact that it covers three basic dimensions of man :

(i) What he does? (ii) Why he does? (iii) Through what means he seeks to accomplish goal.

FUNCTIONS OF PERSONNEL MANAGER

The status of Personal managers differs from unit to unit and industry to industry. They are called by various designations in companies. They perform a number of functions for the achievement of organizations objectives. Broadly speaking, the functions of personnel managers fall in the these areas :

1. Organisational Planning and Development 2. Staffing 3. Training and Development 4. Wage and salary Administration 5. Motivation 6. Employee Services 7. Employee record 8. labour Relations 9. Personnel Research

1. ORGANISATIONAL PLANNING AND DEVELOPMENT

Process of planning and developing an appropriate organizational structure which will ensure effective work performance, fruitful interpersonal relations and formation of homogeneous, cohesive and interacting informal groups. This major function could be sub divided into:

(a) Determining organizational needs in terms of company's short and long term objectives, technology of production, nature of product, external environment public policy etc.

(b) Planning and designing organizational structure which will permit the achievement of organizational goals.

(c) Designing inter personal relationships which will result into healthy and fruitful inter personal relations and formation of homogeneous, cohesive and effectively interacting informal groups.

2. STAFFING

Process of obtaining and maintaining capable and competent personnel to fill all positions from top management to operative level. This major function could be sub-divided into :

(a) Manpower planning : Process of analyzing company needs of personnel now and in future in view of its short and long term objectives by adopting following measures :

(i) Analyse company manpower requirements in terms of short/long term goals.

(ii) Prepare an inventory of management and other personnel.

(iii) Calculate and forecast turnover.

(iv) Prepare a schedule of manpower needs over reasonable period of time.

(v) Develop job description and job specifications.

(b) Recruitment : Process of attracting qualified and competent personnel

(i) Identify existing sources of applicants.

(ii) Develop new sources of applicants.

(iii) Attract potential applicants in sufficient numbers to permit good selection.

(c) **Selection :** Process of developing selection policies and procedures and

   (i)   Evaluating potential employees in terms of job specifications.

   (ii) Develop application blanks.

   (iii) Develop valid and reliable testing techniques.

   (iv) Develop interviewing techniques.

   (v)   Develop employee referral system.

   (vi) Develop medical examination policy and procedure.

   (vii) Evaluate and select personnel in terms of job specification.

   (viii) Make final recommendations to line management.

   (ix) Send rejection and appointment letters.

(d) **Placement :** Process of placing the employees on the job for which he is most suitable in terms of job requirements, his qualifications and personality needs :

   (i)   Advise line management on placement.

   (ii) Conduct follow up study to determine employee adjustment with the job.

(e) **Induction and orientation** : Process of initiating the employee in the organization and on the job :

   (i)   acquaint the employee with the company personality, philosophy, objectives, policies, career development opportunities, product, market standing, social and community standing,

   (ii) Familiarize the employee with the people with whom he is to interact as peers, superiors and subordinates.

(f) **Transfer :** Process of placing employees where they can be utilized more effectively consistent with their social and psychological needs

   (i)   Develop transfer policies and procedures.

   (ii) Counseling employees and line management on transfers.

   (iii) Evaluate transfer policies and procedure.

(g) **Promotion :** Process of advancing employees to higher positions keeping in view their capabilities, job requirements and personality needs :

   (i)   Develop equitable, fair and consistent promotion policies and procedures.

   (ii) Advise line management and employees on matters relating to promotions.

   (iii) Oversee the implementation of promotion policies and procedures.

(h) **Separation :** Process of severing relations with employees in a congenial manner.

  (i)   Conduct exit interviews.

  (ii) Analyse employee turnover.

  (iii) advise the line management on causes of turnover

3. TRAINING AND DEVELOPMENT

Process of training and developing employees so as to develop their full potential for optimum efficiency in effective job performance. This function subdivided into :

(a) **Operative training :** Process of imparting requisite job skills to operators:

  (i)   Identify training needs of the company.

  (ii) Develop suitable training programmes.

  (iii) Identify operatives who need training and who have the aptitude and motivation to go through the training programme.

  (iv) Held and advise live management in the conduct of training programme.

Conduct of training programme

(b) **Executive Development :** Process of designing suitable executive development programme :

  (i)   identify areas in which executive development is needed.

  (ii) develop programmes of executive development.

  (iii) motivate the executives to develop.

  (iv) design special development programmes for promotables.

  (v)   conduct executive development programmes, enlist the service of specialists, or and utilize the institutional executive development programmes.

  (vi) evaluate the effectiveness of executive development programmes.

4. WAGE AND SALARY ADMINISTRATION

Process of compensating employees adequately, equitably and fairly,

(a) **Job Evaluation :** Process of determining relative worth of jobs :

  (i)   conduct wage and salary surveys.

  (ii) determine wage and salary rates.

    (iii)   operate wage and salary programmes.

    (iv)   evaluate its effectiveness.

(b) Incentive compensation : Process of developing administering and reviewing a system of financial incentives in addition to regular wage payment so as to encourage higher level of efficiency :

    (i)   develop an incentive payment system,

    (ii)   advise line management on its operation

    (iii)   reviewing periodically to evaluate its effectiveness.

(c) Performance appraisal : Process of evaluating employee performance in terms of pre-determined criteria so as to enable objective administration of the system of rewards and punishment, and identification of promotables :

    (i)   Develop performance appraisal policies, techniques and procedures.

    (ii)   Overview consistent use of performance appraisal programme.

    (iii)   assist line management in conducting performance appraisal.

    (iv)   review performance appraisal reports, consolidate and report on them.

    (v)   evaluate the effectiveness of performance appraisal programme.

MOTIVATION

Process of motivating employees so as to secure their integration with the organization and attain the optimum level of efficiency and effectiveness towards the achievement of organizational goals. It covers :

Non-financial incentives : Process of motivating employees by creating conditions as to permit the satisfaction of their social and psychological needs:

(a) Develop policies and programmes to attain the satisfaction of their social and psychological needs while at work. This may include reorganizing the socio technical system, restructuring organizational relationship, reorganization of work, restructuring communication system, pushing down decision levels etc.

(b) Advise and guide line - management in the execution of these policies and programmes.

(c) Conduct morale and attitude surveys

(d) Diagnose the health of human organization

(e) Advise line management on the need, areas and ways and means of improving the morale of human organization.

EMPLOYEE SERVICES

Process of maintaining a healthy and effective human organization.

(a) **Safety :** Process of ensuring physical safety of employees at work:

   (i)   Develop technique, policies and procedures for safety.

   (ii) advise and guide line-management in implementation and operation of safety programme.

   (iii) Train first line supervisors and workers in safe practices.

   (iv) Investigate accidents and collect statistics.

   (v)   Evaluate the effectiveness of safety programme.

(b) **Employee Counseling :** Process of counseling and helping the employees in solving their work and non work problems :

   (i)   Motivate employees to seek counsel in the solution of work / non-work problems.

   (ii) Provide counsel and help.

   (iii) Advise line management on the general nature of problems facing the employees.

(c) **Medical Services :** Process of providing medical and health services to employees

   (i)   Conduct periodical medical check ups.

   (ii) Advise employees on hygienic and preventive measures.

   (iii) Render curative and preventive medical assistance and facilities.

   (iv) Advise line management on employees health.

(d) **Recreation, Canteen and other welfare programmes :** Process of providing recreational, canteen, educational and other facilities so as to make company employment attractive :

   (i)   Develop suitable policies, programmes and facilities.

   (ii) Administer these programmes.

   (iii) Evaluate the effectiveness of these programmes.

Employee Record : Process of maintaining up-to-date and complete employees records :

   (a)   Process of collecting information relating to personnel qualification, special aptitude, results of employment, testing, job performance, leave, promotions, rewards and punishment etc.

   (b)   Process of analyzing employees records for preparing employees and talent inventory.

(c) Process of developing information needed for decisions relating to employees transfer, promotion, leave, merit, increases etc.

(d) **Labour Relations :** Process of maintaining healthy and peaceful union management relations :

(a) Grievance Handling : Process of redressing grievances :

(i) design grievance procedure and machinery.

(ii) evaluate the effectiveness of grievance procedure and machinery.

(iii) analyse the nature of grievances.

(iv) identify areas of dissatisfaction and use them as guide in policy making.

(v) advise management on the areas of dissatisfaction.

(b) Discipline : process of developing policies, rules and procedures, relating to employees conduct and behaviour and ensure their observance :

(i) develop policies, rules and procedures to maintain discipline.

(ii) develop a system of reward and punishment to reward the observance and punish violation of discipline.

(iii) analyse the nature and cause of indiscipline.

(iv) advise line management.

(c) Compliance with labour laws : Process of observance and compliance with labour laws.

(i) acquire knowledge of labour laws applicable to the company

(ii) acquaint line management with relevant labour laws and help them in their compliance.

(iii) overview compliance.

(d) Collective Bargaining : Process of negotiating and entering into agreement with labour union on wages, working conditions, employment relationship etc.

(i) Identify areas of disagreement between management and labour.

(ii) Identify areas of collective bargaining.

(iii) Collect date and information inside and outside the company.

(iv) Negotiate or help line management in negotiations.

(v) Interpret and administer the agreements.

PERSONNEL RESEARCH

Process of evaluating the effectiveness of personnel programmes, policies and procedures and developing more appropriate ones:

(a) Conduct morale and attitude surveys.

(b) Collect data relating to productivity, quality, wage, grievances, absenteeism, turnover, strikes, accidents and other indices of operational effectiveness of personnel programmes, policies and procedures.

(c) Report to line management on findings.

(d) Advise line management on the need, areas and directions of change.

(e) Develop more appropriate personnel programmes policies and procedures.

PLACE IN THE ORGANISATION

All decisions, whether, they relate to marketing, finance, production research and development or quality control aspect of organizational activity, have a human aspect and therefore a personnel aspect. Personnel management is thus all pervasive (tending to spread), transcends (go beyond) all other managerial functions and is ubiquitous (present everywhere). In this sense, every manager is unavoidably a personnel manager. The Personal department is a specialized department at the executive level whose function is to encourage, advise and assist line management executives to adopt point of view, develop policies and methods, and apply skills which will release the productive energies of all supervisors and employees.

Challenges faced by personnel managers are varied like obtaining and maintaining the right person, for the right job, at the right time; continuous training and development of employees to cope up with the frequent changes in technology. The multiplicity, multisation and politicization of trade unions poses a major challenge to the personnel manager.

An effective personnel department, with top management's support should work positively to solve these day-to-day problems.

PLANNING AND PERSONNEL SELECTION

Introduction : Management of human resources in contemporary times is highly complex, sophisticated and difficult. The momentous changes taking place in education, science, technology, management, business and society have rendered many age old concepts and stereotypes obsolete. The human resources planner has to have some understanding of the new perspectives both with regard to the challenges and the changing attitudes of people.

Manpower Management : Manpower means people. It means workers. It refers to the adult and working human beings in modern societies the men and women who employ others or are employed in business, industry and government. It embraces women power, to repeat a widely circulated phase. Manpower includes both employers and employees.

Manpower Planning : As in case of any other functional area like marketing, production or finance, the personnel department's work has also to be planned. Planning in the personnel area is mainly concerned with crystallizing from where the right type of people can be secured for future anticipated vacancies. Manpower planning may be defined as a strategy for the acquisition, utilization, improvement and preservation of an enterprise's human resources.

As Vetter defines : The process by which management determines how the organization should move from its current manpower position to its desired manpower position. Through planning management strives to have the right number and the right kinds of people, at the right place, at the right time, doing things which result in both the organization and the individual receiving maximum long run benefit.

MANPOWER PLANNING STEPS

The need to anticipate and provide for future manpower requirements has made manpower planning vital function today in the area of staffing of the personnel function. In large organizations, where a personnel department exists, this function is naturally performed by such dept. as a staff function. Systematic manpower planning has not yet become really popular even in advanced countries such as USA and UK, being practiced there only by a few huge companies in large scale industries such as petroleum and chemicals.

Manpower planning can basically be done by following three steps :

First Step : Determine the period for forecasting requirements of manpower in the future (i.e. requirements at the end of first year, second year, third year, fourth year, fifth year etc.) and forecast the manpower required at the end of such period.

Second Step : From the number available at the commencement of the period, deduct the expected wastage, through deaths, resignations, retirements and discharges. This would give the manpower available from existing staff at the end of the period concerned. A comparison of the figures arrived at step first and second should indicate shortages or surpluses in manpower requirements.

Third Step

(a) In case of shortages decide how much shortages, are to be met (i.e., whether through fresh recruitment and/or promotions from within) and whether any training of developmental facilities would be required for this purpose.

(b) If surpluses are anticipated, decide how these surpluses will be dealt with like through early retirements, discharges or layoff.

PERSONNEL SELECTION

The promotion of productive efficiency through an effective utilization of men and machines is one of the primary objectives of personnel administration and the attainment of this objective largely is contingent upon the function of finding and placing the right man on the right job at the right time in the right place. In order is avoid the pitfalls of wrong selection and placement, it is necessary to adopt the principles of scientific selection procedures. The function of scientific selection is one of the most important functions of personnel administration and this encompasses the following subfunctions:

(A) Determining the nature of the job to be filled in

This is the first stage in the process of placing the right man on the right job at the right time in the right place, through the adoption of a scientific selection procedure. It is essential for the personnel executive to find out the specific nature of the job to be filled before initiating the process of selection.

(B) Determining the nature of personnel required

After the nature of the job is determined, the characteristics of the manpower required to fill the job assume prominence and also the number of employees that must the procured.

The personal characteristics of job specification have to be determined and including. (a) Physical specifications (b) Mental specifications (c) Emotional and social specifications (d) Behavioural specifications.

(C) Determining the nature of source of Recruitment

After obtaining information regarding job requirements and manpower requirements and before instituting the selection procedure, it will be necessary to find out the nature of requirement. The sources of manpower supply are many and varied and the company must know what and where these are, in order to fill their personnel needs. Some organizations have established recruitment policies. It may be a company's policy to hire only the relatives of the employees and one Indian Textile mill has a hierarchy of preferences. First they prefer wives or husbands, then sons or daughters, then cousins and then other close relations. This mill has a policy of intimating its employees of the vacancy or potential vacancy and of inviting employees to recommend a person who must be his relative. For this purpose, the firm uses on employee Recommendation Form. Many companies have a policy of hiring friends and relatives.

There are a variety of sources of recruitment for manual, clerical, sales, professional, technical and managerial personnel and the following sources of recruitment are widely used :

1. **Direct hiring :** This has reference to those who come to the door of the company looking for employment. Those firms which have a good reputation regarding wages, working conditions and other facilities attract a good number of people, from whom the company may think of selecting some.

2. **Friends and relatives :** As already pointed out, friends and relatives of employees are another good source of supply. Often the employees may tell his friend or relative that his company is hiring people and so the word spreads. Those who depend on this source declare that there is a high degree of morale and loyalty and good group relationships and friendliness among the employees. The work atmosphere appears congenial and even enjoyable.

3. **Advertisements in newspaper :** This is one of the most commonly employed source. The corporation needing manpower to fill certain job, advertise the available job, likely pay, duties and responsibilities of the job, job specification and also the man specification in a newspaper, magazine or journal and invites applications. But the advertisements should be carefully written giving all relevant data, and they must attract only the right type of people with right qualifications. If the ad. is vague and general, it may attract hundreds or even thousands of applications, rendering the process of screening time consuming and costly. The ad. must clearly state the educational qualifications, experience and skills necessary to do the job to be filled, with a view to discouraging the unsuitable candidates from applying.

4. **Unions :** This is another source of recruitment and in some industries such as the building, trades unions have provided the employer the necessary number of workers. The union may also advise the worker where he can find a job. If its member loses his job in one company for some reason, the union may find him a job in another company.

5. **Public employment agencies :** The United States Employment Services or "USES' as it is popularly known, has a network of several state employment services, which act as public employment agencies. In the early year they were not found every useful, but they were, after the passage of the social security Act, which provided unemployment compensation to the unemployed for a short time till they found another job. In order to obtain this compensation, the unemployed had to register with USES. In India, Employment Exchanges are assisting the unemployment to find at the jobs.

6. **Private employment agencies :** Are another important source, which many companies cultivate for hiring purposes. The firms can feed suitable men though these private employment agencies. Usually they specialize in supplying certain types of personnel and some specialize in clerical and secretarial personnel and others in executives, accountants, engineers,

economists, salesmen, dieticians, top executives etc. They usually charge a fee for their services and they may collect the fee either from the applicant or the employer or from the both. They also provide employment counseling and guidance services, resume service and other services to the people seeking employment.

7. **Employees' or Trade Association :** Meetings, conference, seminars and other social functions organized by these associations are another source, through which the firm try to recruit the needed people. A few come to these conference looking for a change in positions and some others attend to recruit the right man.

8. **Professional Associations and Journals :** Are yet another source for finding some professional and technical people. Advertising in the journal will bring in good response. Again the seminars, symposia meetings, conferences and other functions sponsored by the professional associations provide opportunities to recruit professional personnel. Though the journals, the employers can identify who contribute articles and papers and who may be tapped.

9. **Schools and colleges :** Recruitment from educational institutions is a popular practices of thousand of firms. Many colleges have their own placement office, through which students find their jobs. The placement office is in touch with various firms and display employment opportunities on its bulletin board for the benefit of students who may apply for some job. The companies usually send their recruiters to college or university campus to interview candidates.

10. **From other firms :** Recruiting personal from other firms is a popular practice. There are corporations, which have made a name for themselves in training and developing people, particularly executives.

11. **Management consultants :** Specialized executives selection services are offered to the companies by many consultants. Executives services is really big fascinating business. In our country there are many consultancy firms which offer their services to the companies.

12. **Radio and television :** Through radio and TV many companies recruit their employee and particularly when the labour market is very tight, those source of recruitment are used much.

13. **Internal sources :** To many organizations, internal source is the main source of recruitment. Through transfer, promotions and demotions, the firms try to fill their personnel needs. Recruitment from within policy is a common policy with many organizations. This policy has many merits (a) it improves the morale of the men (b) it promotes loyalty among men to the organization (c) the employer has tried the familiar employees to evaluate.

But the problem with this policy is that it leads to the danger of "inbreeding". People, when they work for an organization develop attitudes and notions that are molded by the organization and the danger here is the lock of new initiative ideas and view. Growth and development of any organization needs certain amount of exposure to the new ideas and initiative that may come from people who come from outside the organization. Another danger is that internal source may dry up. It may not be possible to find a certain type of skilled labour within the organization. Another problem is that promotion based on seniority principle is inherent in this policy of recruitment from within. Seniority alone should not be the criteria for promotion.

14. **Other services :** In some areas, churches, fraternal organizations lodges serve as effective employment agencies. There are some institutions, where deaf, dumb and blind workers are available for employment if you need them.

D. SELECTION PROCESS

The selection procedure must take into consideration the public policy and operate within the framework of the provisions of the state.

Generally, a good selection process will include the following steps, which act as a sequence of obstacles to be surmounted over come by the candidates seeking employment. Preliminary interviews, Application forms, Reference letters, Group discussions, Interviews, Tests, Physical Examination, Selection and Placement, Induction, Follow up.

1. **Preliminary interviews :** Preliminary interviews, which may be conducted by a junior executive in personnel dept., is the first hurdle which a potential candidate has to overcome and it is possible to screen and eliminate unsuitable candidates. Preliminary interview provides the first opportunity for the candidate to know about the company and the job and whether he is suitable or not, the personnel executive must create a good impression of the company on the candidate.

2. **Application forms :** Application forms are widely used everywhere and constitute one of the fundamental media through which information is gathered about the applicant. Most companies design different application forms for different types of employees mangers, supervisors, employees etc.

3. **Reference :** Generally, the applicant will be asked to give two or three references preferably former employers, or personal acquaintances or friends or professors or famous persons, and the applicant will give the names of those who may speak well of him. It is possible to get impartial assessment or evaluation of the applicant from some of the referees.

4. **Group discussion :** Next in the sequence of obstacles for the applicant is the group discussion. It is a method where groups of the job applicants are brought around a conference table and they are either given a case study subject for discussion. It is for the group to analyze, discuss, find solutions, and articulate their views and they are being observed by a selection panel who judge the group discussion on the basis of such activities as : (a) Initiating the discussion, (b) Explaining the problem, (c) Providing information, (d) Clarifying issues, e. Influencing others, (f) Summarizing, (g) Speaking effectively, (h) Mediating arguments among the participants.

5. **Interviews :** Interviewing is the most universally used tool in any selection procedure and interviews are designed to serve in the important areas of employment, training, human relations and labour relations.

Invitation for the Interview : The letter inviting the applicant for the interview must be sent at least three or four weeks ahead of the time of interview, so that the applicant may plan his trip. The interview letter must also state the place of interview and time of interview.

Place of interview : The place of interview in the employment office must be neat and god looking and it must create a good impression on the applicants.

Time of interview : Some companies stagger the interview time in such a way that no candidate have to wait for more than five minutes. This will call for careful planning of interview time, but, this assist the candidates and also the interviewers.

Some Interview rules : Interviewers should observe the rules : (a) Be courteous to the interviewee and let him feel at home. (b) Effective listening. (c) Do not ask leading or tricky questions. (d) Never argue or interpret or change the subject. (e) Use simple language for asking questions. (f) Be tactful in asking direct and personal questions (g) Encourage the candidate to talk. (h) Try to get relevant information. (i) Respect the interest of the candidate (j) Answer candidate's questions. (k) Lead the interview to conclusion.

Training for Interviewers : To observe all these rules, the interviewers must plan the interview and the questions he is going to ask with particular care. Many companies train their interviewers. The interviewers may be asked to beware of their first impressions, hasty inferences, bias, prejudices, likes and dislikes.

Purpose of Interview : The primary purpose of the interview is to get a complete and correct picture gained through application forms. It measures the ability of the candidate to speak and present his views, his sociability, poise, appearance etc.

TYPES OF INTERVIEWS

There are various types of interviews employed by the companies. In India, most of the interviews may be oral and informal. The different types of interviews are :

(i) Informal Interview (ii) Formal Interview (iii) Planned Interview (iv) Patterned Interview (v) Nondirective Interview (vi) Depth Interview (vii) Stress Interview (viii) Group Interview (ix) Panel Interview

- **(i) Informal Interview :** An informal Interview is an oral interview and may take place anywhere. The employee or the manager or the personnel man may ask a few almost inconsequential questions like name, place of birth, names of relatives etc., either in their respective offices or anywhere outside the plant or company. It is not planned and nobody prepares for it. This interview is used widely when the labour market is tight and when you need workers badly.

- **(ii) Formal interview :** Formal interviews may be held in the employment office by the employment officer in a more formal atmosphere, with the help of well structured questions, the time and place of the interview will be stipulated by the employment office.

- **(iii) Planned interview :** It is a formal interview, which is carefully planned. Here the interviewer may have a plan of action worked out in his own mind and he knows how much time he is going to devote to each candidate, what type of information he is seeking and what he proposes to give, how to open interview and how to close the interview and how to conduct the interview. He may use the plan with some amount of flexibility. He may deviate from his plan, but he knows what he is doing and he can come back to his original plan and continue the interview.

- **(iv) Patterned interview :** A patterned interview is also a planned interview, but it is more carefully pre-planned to a high degree of accuracy, precision and exactitude. With the help of job and many specifications, a list of questions and areas will be carefully prepared, and it will act as the interviewer's guide. But the interviewer's guide may be supplemented to gather any other significant information and this is dependent upon the skill of the interviewer.

- **(v) Non-directive interview :** Non-directive interview or unstructured interview is designed to let the interviewee speak his mind freely. The interviewer has no formal or direct questions, but his all attention to the candidate. He encourage the candidate to talk by a little prodding whenever he is silent, e.g., "Mr. Ray, please tell us about yourself after your graduated from high school". The idea is to give the candidate complete freedom to "self himself", without the encumbrances of the interviewer's question. But the interviewer must be of a higher caliber and must guide and relate the information given by the applicant to the objective of the interview.

(vi) **Depth interview :** It is designed to intensely examine the candidate's background and thinking and to go into considerable detail on particular subjects of an important nature and of special interest to the candidate for example, if the candidate says that he is interested in tennis, a series of questions may be asked to test the depth of understanding and interest of the candidate. These probing questions must be asked with tack and through exhaustive analysis, it is possible to get a good picture of the candidate.

(vii) **Stress interview :** It is designed to test the candidate and his conduct and behaviour by putting him under conditions of stress and strain. The interviewer may start with "Mr. Joseph, we do not think your qualifications and experience are adequate for this position, and watch the reaction of the candidate. A good candidate will not yield, on the contrary he may substantiate why he is qualified to handle the job."

(viii) **Group interview :** It is designed to save busy executive's time and to see how the candidates may be brought together in the employment office and they may be interviewed.

(ix) **Panel Interview :** A panel or interviewing board or selection committee may interview the candidate, usually in the case of supervisory and managerial positions. This type of interview pools the collective judgment and wisdom of the panel in the assessment of the candidate and also in questioning the faculties of the candidate.

LIMITATIONS OF THE INTERVIEW

(a) Halo effect (b) Stereotyping (c) Subjective Elements

Tests are used to gather some more accurate information about the candidate and thus, they supplement the other selection techniques and they depend upon correct construction and interpretation. There are many hazards when tests are used as sole criterion for selection. But tests are found useful as one of selection techniques. Some common test used are : (a) Attitude Tests (b) Intelligence Tests (c) Mechanical Tests (d) Trade Tests (e) Character tests (f) Achievement tests (g) Combination tests (h) Personality tests (i) Interest tests. (j) Various tailor made tests.

6. Physical Examination

It is common to give the prospective employee a through medical checkup, known as the pre-employment physical examination. If the employee's health is found satisfactory, a job offer is made. This is just another selection hurdle that has to be surmounted by the candidate.

7. Selection and Placement

Interpreting the findings made though interviews, application blanks, references, personal observation and physical examination is the next process of selection, although at each of the selection hurdles, the selection process is taking place. On the basis of the findings, decisions to select is made. The responsibility for such decision should rest with the line executives, same in the case of recruitment for the personnel division.

8. Induction

The selection procedure does not end once the selection is made, because the new employee has to be inducted into organization. The purpose of induction is to educate the employee and provide orientation.

9. Follow Up

The objective of a follow up is to see whether the right man has been placed in the right job or there has been a mistake. This also provides an opportunity for the supervisors or the manager to assess the contribution of the employee and make suggestions, if necessary, to improve his performance.

Scientific selection is one of the most important principles of scientific management and there is enough knowledge and experience to prove that these selection techniques are useful in the recruitment, selection and placement process.

TRAINING NEEDS - METHODS AND EVALUATION

Training and Management Development

One of the central and legitimate purposes of modern management and business is the development of people by providing the right environment where the individual may grow to his fullest stature and realize his fullest potentialities. Development of people is a specialized function and is one of the fundamental operating functions of personnel administration though the responsibility for the training and development of people rests with the line management. In the development and growth of people lies the organizational growth and progress.

The great issue facing not only the industrially developed nations but also the fast developing nation is the problem of change. Combating problem of change and adapting to change, is the primary concern of thinking individuals. Change that is induced by business and industry through science and technology, demands rapid individual and social adjustment and it renders obsolete products and processes, skills and attitudes and with them men and jobs. Meeting this challenge of change is a necessary responsibility of management.

In developing counties like our, there is a great demand for trained personnel staff levels, and this is created by tremendous growth in number, size and complexity of industrial and business organizations to conduct their own programmes and an increasing number of firms are recognizing the training needs of their personnel and are sponsoring a variety of programmes to meet these needs. The basic objective of the personnel executive is to assist the entire organization from top to bottom in bringing about an improvement in knowledge, skill, habits and attitudes that will ultimately express itself productively in work and constructively in human relations. The essential purpose of training is to develop that knowledge and those skills and attitudes, which contributes to the welfare of the company and employees. Further, all training programmes aim at making the employees more effective and productive on their present jobs and increasing their potential on higher level jobs.

TRAINING NEEDS AND OBJECTIVES

Training is no panacea to all manpower problems and so it is necessary to establish where it would be useful and where it would meet a problematic situation. When a college graduate a hired, there is a need to give him some sort of on-the-job training so that he may perform effective and productive. When a new machine is brought into the organization, the people who are going to work on this machine may need some training. There are many industrial and business problems or situations when training and development programmes will be of great use. In modern business, there are various changes job changes, organization changes, method changes, changes in personnel, changes in the volume of business etc., that are constantly taking place and that necessitates modification of understanding, attitudes and skills on the part of the personnel. They create needs that may be met by training and only needs that can be met by training are training needs.

The following formula is suggested to indicate the specific training needs:

Training needs = Job Requirements – Employee's preset job skills.

This indicate that there is a need for careful job analysis and job specification and a careful analysis of the individual employees. A survey of overall training needs of an organization may be conducted periodically with the active support of top management. There may be few persons, who, may think it is just another personnel gimmick and it is of no practical value to the organization at all. They must be converted by pointing out the specific results of training. And that training is a practical necessity which results in the reduction of waste and spoilage, improvement of methods, reduction of absenteeism and labour turnover, reduction of learning time etc.

METHODS OF DETERMINING TRAINING NEEDS

1. Analysis of an activity (Process, job, operation)

2. Analysis of problems

3. Analysis of behaviour

4. Analysis of an organization

5. Appraisal of performance

6. Brainstorming

7. Committee

8. Comparison

9. Conference

10. Consultants

11. Counseling

12. Informal talks.

13. Interviews

14. Observation

15. Self Analysis

16. Skills-inventory

17. Studies

18. Surveys

19 Tests

20. Questionnaire.

Workshop

Training objectives and policy formulation must be able to provide answer to the questions : (a) What do you want and hope to accomplish through training? (b) Whose responsibility is training? (c) In training to be formal or informal? (d) What are your priorities in training? (e) What types of training do you need? (f) When and where shall training be given? (g) shall training be continuous or continual? (h) What outside agencies will be associated with training?

Advantages of Training : Training is the act of increasing the knowledge and skill of an employee for doing a particular job. The major advantages of training are :

1. Increased productivity.

2. Heightened morale.

3. Reduced supervision.

4. Reduced accidents.

5. Increased knowledge.

6. Improved skills.

7. Decreased fatigue and efforts.

8. Increased organizational stability and flexibitility.

9. Improvement in quality, communication.

PRINCIPLES OF TRAINING

1. **Motivation :** An employee who is highly motivated, learns new skills or acquire job knowledge.

2. **Progress Report :** The trainee must be given adequate information as to the progress and effectiveness of learning.

3. **Reinforcement :** When skills are learned the effect should be reinforced by means of rewards and punishment, promotions, pay increases.

4. **Practice :** To effectively acquire skill, knowledge or attitude, active participation is essential

5. **Whole v/s Part :** The work should be divided into meaningful parts. The trainee then has to put the parts together to make the whole.

6. **Individual differences :** It is apparent that individuals vary in intelligence and aptitude. The training procedure has to take account of these.

System of Training : Generally an establishment has following systems of training :

1. **On the job Training :** A worker is given training directly on the job which he is supposed to perform in industry. The trainer makes him work on the job and inparts training through actual work at the machine. The time span depends on the skill required for the job.

2. **Vestibule Training :** Under this system a miniature workshop is set apart and the worker is trained for the first month or so in this workshop prior to taking him on the actual job. This system is not common due to shortage of machinery in Indian industry ad also as this system proves expensive.

3. **Apprenticeship programme :** Apart from the requirements under Apprenticeship Act, various industries have their own Apprenticeship programmes wherein workers are employed on a small stipend to learn the job first and after satisfactorily completing the training they are absorbed in the industry.

4. **Special Courses :** There is a growing awareness among at the industries towards the need for special training courses for all level of employees. This awareness has come as a result of advantages that have been actually seen and felt by the managers of industry.

A good trainer or instructor must : (a) Know the job or subject he is attempting (b) Have the aptitude and abilities to teach (c) Want to teach.

(d) Have a pleasing personality and capacity for (e) Have a knowledge of teaching principles and.

Supervisory Training : The supervision is the main communication link between managers4 and employees and he is the one who has to get things done through his men. He has a very important role to play.

Management is interested in supervisory training and development programmes for a variety of reasons.

(a) As a company expands it needs more supervisors, hence there is a need to augment the source of supply.

(b) Supervisory development programmes are often the basis of executive development programme.

(c) Each dept. needs a good supervisor as its head.

(d) Good supervisor is responsible for good team spirit and team work.

(e) Top management realizes that there is a need to keep the supervisor in good shape all the times.

Training methods for Supervisors : The techniques and methods available for companies to train supervisors and foremen are many and they are discussed below :

(i)	(a) Job Rotation	(d)	Interplant and industrial visit
(ii)	(b) Reading material	(e)	Individual assignment with Dept.
(iii)	(c) Staff meetings	(f)	Participation in the work of other Deptts.
(iv)	(g) Problem solving	(h)	Lectures and Teaching
(v)	(i) Role playing	(j)	Case studies
(vi)	(k) Conferences	(l)	Programmed Instructions

Objectives of Executive Development

(i) More efficient position performance

(ii) More economic methods of work

(iii) More harmonious team work

(iv) Greater morale and job satisfaction

(v) Quicker adaptation to changing conditions

(vi) Grooming for higher position

(vii) Sharpening of appetite for self development

Techniques for management development

(i)   Coaching and counseling

(ii) Position Rotation

(iii) Understudy

(iv) Role playing

(v)   Transitory/Anticipatory Experience

(vi) Self improvement programme

(vii) University management programmes

Evaluation of Training : The criterion is suggested for evaluation of training programmes are (a) Increased output, (b) Reduced time to turn out a unit of production, (c) Reduced training time, (d) Reduction in scrap, breakage and supplies used, (e) Improvement in quality of output, (f) Reduction in absenteeism, grievances, turnover and accidents, (g) Improvement in morale, (h) Reduction in overhead.

Training is provided to meet a need or to solve a problem. If safety training is offered to prevent accidents, the follow up must evaluate whether there have been fewer accidents. The techniques used to evaluate the effectiveness of training programme : are Tests, Attitude Surveys, Cost Accounting, Checklists, Learning curves, Employee Appraisal.

PERSONNEL MANAGEMENT ON FARM

I. Introduction

Labour efficiency in agriculture refers to the amount of productive work accomplished per man on the farm per unit of time. Inefficient labour results in low production, which in turn means low wages for the labour. The labour efficiency also can be expressed as labour input to its output.

On Indian farms, land is limited, i.e., farms are small in size with extremely limited capital and organizational or managerial ability is low. Labour on the other hand is abundant. The resource availability on the farms is thus imbalanced leading to a low production, which results in low returns to the farm business and low farm family labour earnings.

II. Classification of farm labour

Farm labour can be classified into four categories

(i)   Farm manager's labour

(ii) Farm family's labour

(iii) Permanently hired labour

(iv) Casual hired labour.

The first three categories constitute permanent labour force available on the farm and is a fixed resource due to a general lack of mobility. Fourth category is a variable input and can be hired when needed.

III. Composition of farm labour

Agricultural labour are those persons who are engaged as hired labourers in agricultural operations for 50% of the total number of days worked by them and the agricultural wages are the main source of income.

Indian farm labours consists of skilled, semi-skilled and unskilled workers.

(i) **Skilled labour :** Specialized and trained labour for specific job is known as skilled labour viz., drivers, mechanics etc. Some of them may not be needed too often, but their ready at hand availability is however, essential. Non-availability of them generally upset the farm operations. Wages of the skilled workers are always higher than those of the other categories of labour.

(ii) **Semi-skilled labour :** Semi-skilled labour does the job, which cannot be taken up by ordinary labourer, but at same time does not require any elaborate training. The wages of such workers are a little higher than those of the ordinary unskilled labourers.

(iii) **Unskilled labour :** It is ordinary labour employed for manual work, which does not need any training of specialized nature. Unskilled labour is generally engaged in fieldwork as cattle attendants, cotton pickers etc.

IV. Special features of Indian farm labour

1. They are large in number
2. They are mostly unskilled
3. They lack staying power
4. Majority of them are under debit and lack bargaining power
5. Their payments of wages are low, many times seasonal and in kind.
6. Their hours of work are long and irregular
7. Employment opportunities are meagre, seasonal and uncertain.

V. Methods of improving labour efficiency

Improvement of efficiency of farm labour in India is very much essential. Some of the methods, which have been found to be useful in improving the labour efficiency, are: Enlarging the size of the farm business, Planning labour distribution, Improving the field and farm layout, Improving the farm buildings, Labour management, Farm work simplification

1. **Enlarging the farm business :** By enlarging the size of the farm business through (i) expansion of land area if possible through leasing or purchasing and (ii) adding more labour intensive enterprises to intensify the farm business, the efficiency of farm labour can be improved. The labour utilization can be improved by increasing the size of the farm business through improved production techniques.

2. **Planning labour distribution :** Farm work should be planned so that there is employment through each working day and through each season of the year. This can be achieved by : (a) Enterprise combination and (b) mixed farming

    (a) **Enterprise combination :** There are some peak workload periods and a few slack months on Indian farm in case of labour employment. This leads to low level of labour efficiency. If different crop enterprises are planned such that as far as possible peaks and slacks are reduced, labour efficiency will improve. The crop calendar for this purpose can be prepared. A well-adjusted diversified farming is very conducive to improve the labour efficiency on the Indian farms.

    (b) **Mixed farming :** Mixed farm is a better solution for labour efficiency improvement. Addition of dairy cattle to crop farming can absorb surplus labour in slack season. Poultry and piggery are other enterprises that can be profitably included in the production programmes to expand employment in the slack periods.

3. **Improvements in field and farm layout :** Large number of operational holdings remains small and fragmented even after consideration of holdings. This makes scientific layout of farms difficult. Which results into labour wastage in different operations.

    An efficient layout of the farm consists of straight roads, bunds and water channels with a rectangular shape. This economics the movement of labour and other equipment from field to field and increases efficiency in the performance of the operations.

4. **Location of the farm buildings :** Much of the time is wasted in moving from village to farm. Stores, bullocks/cattle sheds and other buildings should be located on the farm so that distance of individual field from the buildings is the minimum. The buildings should be in the centre of the farm or near to main road or highway.

5. **Management of labour :** An appraisal of total labour needs on farm guides in the better management of farm.

    The following specific steps can go a long way in increasing the labour efficiency.

(a) **Planning the works :** Efficient labour management must be adopted by way of :

    (i) Making a labour calendar to keep down the peak labour requirements.

    (ii) Preparing a plan of works and duty list well in advance.

    (iii) Checking tools and implements in advance of their use.

    (iv) Assigning work according to capabilities/aptitudes/liking of worker.

    (v) Setting up work standards for different types of works.

(b) **Incentives :** Labour is most efficient when incentives for work are strong. Better housing, recreation and health facilities are provided. Produce sharing or extra bonus fro better and more works are provided. Wages be paid in time and genuine debit needs may be met. All workers need to be given a human treatment and sympathy in their difficulties.

6. **Farm work simplification :** Work simplification deals with work methods. By improving the work methods more efficient use of labour and other resources can be achieved and more and higher quality work is thus accomplished in less time and with less energy. It examines each part of work carefully to determine

    (i) necessity of job works could either omitted or combined

    (ii) simplified methods in order to save time and money

    (iii) supplies/inputs located centrally

    (iv) tools and machinery are of efficient type and size

    (v) rearrangement of building, gates, doors to avoid the wastage of time.

ROLE OF ENTREPRENEUR

The role that the entrepreneur plays consists of coordinating and correlating the other factors of production. He starts the work, organizes and supervises it. He undertakes to remunerate all the factors of production to pay rent to the landlord. Interest on capital and wages to the labour and pay them in advance of the sale of goods. Whatever may be the outcome, he must be prepared to asset it. He thus takes the final responsibility of the business.

If he has anticipated the consumer's whishes right and interpreted them accurately, he is amply rewarded. The (i) organizing and (ii) risk taking of uncertainty bearing as it is sometimes called are the two chief functions of the modern entrepreneurs.

The entrepreneur is the innovator. Innovation by the entrepreneur implies a variety of things. It may mean, the introduction of new methods of production or an improvement in the old method. It may consist of the introduction of new commodity or a new model or make of an old product.

Innovation may refer to the discovery of new materials; fresh sources of old materials of new users for materials of final goods. It also includes the opening of new markets. Innovation may take the form of new techniques in the way of administration, finance, marketing f human relations inside the business and public relations outside, i.e., with suppliers of materials and customers of products. It is involved finally when new forms of business organisation are instituted such as chain stores, the merger of several establishment of a monopolistic combination among producers.

It will be easily understood that uncertainty is inherent in the making of the decisions like those enumerated above and also in any innovations that may be adopted. The all-embracing function that the entrepreneur performs is, therefore, that of uncertainty bearing.

FINANCIAL MANAGEMENT

1. INTRODUCTION

Accounting is the administration of the economic activity of the business entity. It covers within its scope the maintenance of records and books, which would generate useful information pertaining to the economic activity of the business. This information is available from the account books. From these books of accounts the financial statement are prepared at a periodical interval. This process usually is a post-mortem of the economic activity, which is referred to as financial accounting, which speaks invariably about the PAST but nothing about the future. In the business world of today there is an emphasis on the need of accounting information not only about the past but also about the present and the future. This is known as the shift from the conventional book-keeping or Financial Accounting system which generates information relating o the economic activity of a business entity concerning the future.

2. THE PURPOSE OF THE ACCOUNTING SYSTEM

(1) To facilitate the determination of net profit/loss as well as the valuation of assets and liabilities, which is finally placed through the Financial statements. This is possible through the maintenance of records pertaining to the sales, other incomes, inventories and expenses/costs. Even the values of the assets and liabilities are available from these records. Hence the records and books of accounts are to be maintained to serve this purpose.

(2) To generate useful data to assist management in the areas, viz., product pricing, project analysis and planning and cost analysis.

(3) Performs the useful function of planning and controlling business operations, through the techniques viz., cash flow, profitability analysis, ratio analysis, standard costing, budgetary control, marginal costing, etc.

Hence, it may be conducted that the accounting system is a multi-purpose system, the ultimate objective behind which is to enable the management to achieve its objective of maximizing the return of capital employed or investment by maintaining the sustained rate of growth of the business.

3. FINANCIAL STATEMENT

Whatever may be the pattern of accounting information system; it is aimed at preparation of Financial statement. Each business entity is required to prepare a set of the financial statement at a definite periodical interval. The financial statement so prepared include, *inter alia* the Balance Sheet and the Profit and Loss Account. The answer to this question is found in the following point :

(A) Every company (whether public limited or private limited and whether Indian or foreign) is required to prepare these financial statements under the provisions of Companies Act, 1956. These exists a set of provisions in the act, which lays down the form and manner in which these statements are to be prepared, the time-limit within which these statement are to be submitted and so on. Any non-compliance or contravention of these provisions attract severe panel consequences.

(B) Sole-proprietorship concerns, partnership firms. Hindu individual families and association of person) it has now become mandatory to maintain certain accounting systems as a result of the enactment of Taxation Amendment Act, 1975. Accounting to this, any business entity having either a turnover exceeding Rs. 2.50 lakhs or a net profit exceeding Rs. 25,000 during any of the previous accounting years must maintain the books of accounts in the prescribed manner.

Hence, it may be observed that due to various statutory provisions, the maintenance of the books of accounts and the preparation of the Financial statement therefore, has become obligatory in the case of most of the business entities.

4. REPORTING OF THE FINANCIAL STATEMENTS

These statements are meant for the internal reporting and external reporting.

Internal Reporting

This implies the reporting of these statements to the Management. The term Management may encompass the Board of Directors, Managing Directors, Chief Executives, General Managers, various functional directors, etc. as the case may be. The said internal reporting helps the management : To take decision on the policy matters, To do planning for the future, To exercise control, To appraise the performance at various levels of organizations, To motivate the persons to achieve better performance and To ensure proper co-ordination.

In the interest of the organization, timely and accurate reporting of these statements to the management is essential. As a matter of fact, for the purpose of internal reporting, these statements are prepared at the intervals more frequent than what the law requires. For instance, according to the Companies Act, and the Income Tax Act the Balance Sheet has to be prepared monthly and that too, with more details. The exact pattern of the internal reporting would depend upon the philosophy of the management and its expectations from the accounting information system.

External Reporting

(1) **Shareholders :** The shareholders are entitled to receive the profit and loss account and Balance Sheet for the year within six month of the expiry of the accounting year, in accordance with the provisions of the Company Act. These provisions, which are very stringent in nature, are enacted with a view to bridge the communication between the Management of the Company - represented by the Board of Directors and the Ownership - represented by the shareholders, since the ownership has a financial stake in the company, but no chance to participate in the management.

(2) **Financial Institutions/Banks :** Where the funds for the business are produced from the Bank and or/Financial Institutions, they are entitled to receive the financial statements as leaders from their borrowers, since they have staked their funds in the business of the borrowers. Apart from he annual financial statements, such lenders may ask for the financial statement at quarterly/half-yearly intervals, and that too in the form prescribed by them.

These lenders are interested in the assessment of the creditworthiness of the borrowers and to know the manner in which the funds lent by them are employed by the borrower. Further, they are also insuring the repayments of loan and interest.

(3) **Taxation Authorities :** Each business entity is liable of abide by the tax laws, which includes *inter alia*, filling of returns within the prescribed time limits, that too in the prescribed forms and payment of tax dues

within he prescribed time-limit for this purpose, it is essential to prepare the financial statements from the accounting information systems, in the absence of which, it may not be possible to file the statutory. Moreover, the provisions of Income Tax Law regarding the advance payment of Income Tax have made it obligatory for each business entity to compute its taxable income sufficiently in advance and pay the tax thereon accordingly after filing the estimate taxable income within the prescribed time-limit. The contraventions of these provisions attract severe penal consequences. Hence the accounting information system must be so geared as to ensure due compliance with the provisions of the tax laws, so that the financial statements can be readily prepared and the tax ability can be met on the basis of profit/loss revealed by the same.

(4) **Government and Semi-government authorities :** In addition to the taxation authorities, the business entity has to takecare of the requirements of various government and semi-government authorities for whom the financial statements in the form of one or the other are needed.

It is impossible to prepare an exhaustive list of such authorities, still the following names are given as the illustrative list :

(a) Company Law Board and Registrar of Companies under Companies Act.

(b) Director General of Technical Development (DGTD).

(c) State Directorate of Industries.

(d) Chief Controller of Import and Exports for Import/Export Licensing.

(e) Export Promotion Councils incase of the registered exports.

(f) National Sample Survey.

(g) Reserve Bank of India.

(h) Trade Association.

Miscellaneous : There are various other persons/parties that are interested in the financial statements. They include *inter alia* debenture holders a long-term creditors of the company or suppliers who may be interested in knowing the solvency position of the company for the purpose of extending credit or some other concessions. These statements may be needed by the Labour Court and/or other industrial Tribunal with reference to some industrial disputes between the employed and the employees.

5. Understanding Financial Statements

As stated earlier, the financial statements include Income statement and Balance Sheet. The Income Statement is popularly known as Profit and Loss

Account. It is prepared for a given period say for the year ended 31.12.2008. It can be prepared for a shorter period say for the one month, if the management so desires. It reveals the result of the business operations during the particular period, for which it is prepared. This result is expressed in terms of profit/loss. The Balance Sheet is a statement of the assets and liabilities prepared as on a particular date. It is a statement of affairs of the business. There exists certain subtle difference between the income Statement and the Balance Sheet, which may be highlighted as follows :

(A) The Balance Sheet is a statement as on a particular date, hence recognises the assets and liabilities as they stand on that day only while the Income Statement covers all the taken place during the period for which it is prepared and the result is ascertained on the basis of the same. This difference is apparent from the very heading of these two statements. The heading of income statement reads as the Profit and Loss Account for the year ended.../......../2008.

(B) The Income statement reveals the result of this business operations during a particular period, hence it may be called a result-oriented statement while the Balance Sheet reveals the state of affairs or the position of the business, hence it may be called a 'Position-orient' statement, reflecting the financial position of the business.

(C) The Income statement includes the items of Income/Revenue on one hand and the Expenses/Costs on the other hand, the difference between which shows profit/loss. The Balance Sheet, on the contrary, shows the assets and the liabilities, the difference between which indicates the net worth of the business.

Reviewing these differences between these two it may be concluded that both these statements are different from each other, having distinct role of play. Nevertheless, both these statements are essential to carry out the critical appraisal of the business, they are not parallel but supplementary to each other and each one reveals something, which the other one does not. Hence, both of them are needed together by observer or analyst and one without the other cannot convey any useful information.

6. The Income Statement

The Income statement is known as Profit and Loss Account, as explained earlier. It shows the summary of income earned and expenses incurred during the period. It includes :

(a) Sales of products of services.

(b) Income other than sales.

(c) Manufacturing or factory cost.

(d) Office and Administrative cost.

(e) Selling and Distribution cost.

(f) Finance Cost.

(g) Non-operating cost.

BALANCE SHEET ANALYSIS

The main object of farm business analysis is to examine as to (i) how does the business fare at a certain time; (ii) Where are the weaknesses; and (iii) What improvements are possible.

There are some subsidiary objectives too, such as providing background information for farm policies and for getting credit facilities etc.

The following are the three steps or stages of farm business analysis:

(i) Proper recording of accounts and activities,

(ii) Analysis and interpretation of results, and

(iii) Presentation of results.

(i) **Recording of Data :** Proper recording of the data in the relevant columns of suitable record books is very essential. The daily transactions need to be recorded neatly and correctly in the appropriate columns meant for the purpose. Making summaries and analyzing the recorded data becomes very difficult if systematic method of making the entries is not followed and sometimes all the time and effort put in a haphazard record are lost. It is, therefore, necessary to select suitable types of record books.

(ii) **Analysis and Interpretation :** Raw data in the farm financial records kept by farmers will have little value unless they are properly summarized, tabulated and analyzed. The second stage of farm accounting begins with the determination of the proper measures of income and the computation of management and efficiency factors for the farm.

Finally tabulations and charts are made to show the factors, which affect the farm success and failure so that an individual farmer by comparing his performance with these measures or standards may be able to recognize his weak points and take necessary steps for improvements.

ADVANTAGES OF FARM RECORDS AND ACCOUNTS

The various advantages of keeping systematic farm records can be described as under:

1. **Means to higher income :** To obtain higher income, farmers must have exact knowledge about present and potential gross income and operating

costs. The best way to obtain information on present results is to keep records and accounts, in order to:

   (i)   Know the financial status at a point of time.

   (ii)   Know gains and losses over time.

   (iii)   Know the better sources of income and items of cost.

   (iv)   Keep a check on unproductive expenditure.

   (v)   Examine comparative profitability and costs involved for different enterprises.

   (vi)   Locate weak points in the farm organization.

   (vii)   Compare farm efficiency with the farmers operating under comparable farming situations, and

   (viii)   Develop rational short-term and long-term production plans.

2. **Basis for diagnosis and planning :** Diagnosis of management problems is the pre-requisite of sound planning. Records and accounts provide the basic information needed for such a diagnosis.

3. **Way to improve managerial ability of the farmer :** It helps to acquire business habits, which can be of help in taking advantage of changes in the economic environment. The farmer gets a better insight into the working of his business, which helps him in finding out the defects which can be set right by exercising better control and effecting economics. The farmer can avoid mistakes and losses, which would otherwise result due to dependence only on his memory for guidance.

4. **Basis for credit acquisition and management :** Properly kept records and accounts can demonstrate and authenticate the production and income potentials and creditworthiness of the farmer. Thus lending agencies can help the farmer in meeting his credit needs more readily and he can manage his credit utilization properly.

5. **Guide to better home management :** Records and accounts provide information on farm household economy. This is particularly important in India where farm and home management are so closely integrated. Analysis of farm records provides good guides for the allocation of resources between production improvement and immediate family welfare.

6. **Basis of conducting research :** Research requires precise and correct data, which is possible, only if proper records and accounts are maintained on the farms included in the study.

7. **Basis for Government policies :** The farmers need to continuously feed the facts for state and national farm policies such as land policies, price

policies, and crop insurance, etc. Records and accounts are helpful in obtaining the correct data for examining and developing such policies to be sound.

PROBLEMS AND DIFFICULTIES IN FARM ACCOUNTING

1. **Subsistence nature of farming :** Farming as a business is, relatively speaking, a small size operation in India. Farmers cannot engage separately trained accountants for helping them in farm accounting, Subsistence nature of farming does not produce any incentive for keeping the records.

2. **Farming is a laborious work :** Farming requires a lot of physical labour, in addition to mental work of management. In the daily routine, the farmer usually gets exhausted in the evening and does not feel like sitting at the desk to complete records and accounts.

3. **Triple role of Indian farmer :** Indian farmer plays a triple role in running his business, i.e., that of a manager, a financer and a labourer. He needs, therefore, to know his wages as a labourer, his returns to capital and his returns to his management role complex type of records, which would give such information, are difficult to maintain.

4. **Inadequate extension service :** Sufficient numbers of trained specialists in agri-business management are not available who could help farmers maintain accounts of their business.

5. **Illiteracy and lack of business awareness :** The very low level of literacy among the Indian farmers is a hindrance in developing the required level of business awareness on the part of the cultivators and they do not, therefore, realize the need for records and accounts.

6. **Complicated nature of the agriculture business :** It is a biological industry and is always subject to weather and other natural uncertainties. It requires an accounting system, which can handle various complexities involved in the business of farming. Such complicated accounts are difficult to maintain.

7. **Non-availability of suitable farm record books :** Lack of standardized, easy to understand and maintain account books or Performa also stand in the way of willingness of the cultivators to keep records. Standard farm record books need to be developed which may be simple and easy to understand and available in local languages.

8. **Fear of taxation :** Farmers are always afraid of new taxes. They fear that if they maintain records and accounts and their incomes show up high, some sort of tax maybe levied on them.

FARM RECORDS ARE USUALLY OF THE FOLLOWING TYPES

1. Farm inventory
2. Farm financial accounting
3. Farm cost accounting
    (a) Full cost accounting
    (b) Single enterprise accounting

Balance Sheet Analysis

The risk bearing ability is reflected by the balance sheet, the financial statement which indicates the equity. The balance sheet shows much of the ability of the business to weather steems and pick up seed in prosperous times. It shows the fundamental soundness of a business.

Meaning : The balance sheet, also called as the net worth statement, is a summary of the assets and liabilities of a business, together with a statement of owners equity. The owner's equity is also commonly referred to as the net worth of the business.

Components of Balance Sheet

The balance sheet has two characteristics that should be kept clearly in mind.

1. It always refers to a specific date or point in time and
2. It always divided in to three components
    (a) The assets or value of things owned
    (b) The liabilities or amounts owned
    (c) The difference between these two, which is the owner's equity or deficit.

It is this last item, the owner's equity or deficit, which makes the statement of balance.

A balance sheet represents a snapshot of the business at just one instant of time.

A balance sheet be recognized for what it is, a statement of financial position of the farm business as of a certain date. The balance sheet gives the inventory value of property owned, subtracts from thus the total debit owned, and calls the remainder the equity or net ownership of the farmer in his business.

STRUCTURE OF BALANCE SHEET

(a) *Assets* : Are the goods and services, which are owned by the farmer.

(b) *Liabilities* : Are the debts or outstandings which are owned by the farmer.

When assets exceeds liabilities or debts the difference is owner's equity or net worth and placed on the "liabilities side" in order to balance the statement. The farmer is solvent in this case.

When the value of liabilities or debts is greater than the value of assets, the difference is called "net deficit" and the farmer is insolvent to that extent. A net deficit is placed on the Assets side of the balance sheet because it represents the shortage of assets.

BALANCE SHEET - FARM BUSINESS A

Balance sheet as on 1st January 1995

Assets	Rs.	Liabilities and owner's equity	Rs.
Current	**Current**		
Cash, grains, fodder, cattle poultry	11925	Account payable, hardware store, AC payable, tractor, fuel bank loans.	7653
Intermediate	**Intermediate**		
Farm machinery and equipments farm vehicles	8173	Loan for Dead unsecured	5220
Fixed	Nil	Long term owner's equity	7225
Total	20098	Total	20098

NEED FOR PHYSICAL DATA

The information on physical data particularly yields, output, prices, livestock, etc. is required in preparation on balance sheet.

TREND ANALYSIS

As a income statement for one year may not portray a reliable picture of the income which might be expected. Similarly a balance sheet as of a given time may not give a representative picture of a financial situation due to unusual factors affecting production or prices at the time.

It is desirable to have a series of balance sheets for a business to provide a representative financial picture and also to show trends in the financial structure of the business. These were reasons why leaders make a practice of obtaining a current balance sheet with each loan application.

(i) *Ratios and comparative analysis* : A number of ratios have been formed to be important indicators of financial progress and of risk bearing ability.

A classic measure of financial condition used in balance sheet analysis is the current ratio which indicates the current liability of the business.

(a)   Current ratio $= \dfrac{\text{Total current assets}}{\text{Total current liabilities}}$

(b)   Intermediate ratio $= \dfrac{\text{Total current and intermediate assets}}{\text{Total current and intermediate liabilities}}$

The intermediate ratio is used to reflect the intermediate liquidity position of the business.

The long run liquidity position of a business is indicated by the net capital ratio.

(c)   Net capital ratio $= \dfrac{\text{Total assets}}{\text{Total liabilities}}$

Another overall measure of liquidity commonly used is the ratio of total debts to owner's equity.

(d)   Debt to equity ratio $= \dfrac{\text{Total liabilities}}{\text{Owner's equity}}$

The equity is often related to the value of assets the ratio is

(e)   Equity to value ratio $= \dfrac{\text{Owner's equity}}{\text{Value of assets}}$

INCOME STATEMENT ANALYSIS

(1)   **Meaning :** An income statement, also called a profit and loss statement is a summary of receipts and gains during specified period, usually a year, less expenses and losses during the same period, with a net income or a net loss as a result. It is a measure of output and input in terms of values.

(2)   **Receipts :** Receipts are derived from sales of crops, livestock's and livestock products during the year and also from government payments and miscellaneous sources. Any farm product used in home should be valued and also included in receipts.

The objective of the receipt section for the income section is to show as accurately as feasible the gross production of the farm, during the year. This facilitates

(a)   Comparison of given farm with other similarly farms in the area from a management point of view.

(b)   Analyzing the trend of income on the given farm over a period of year Increases are added to and decreases are substracted from gross cash receipts to obtain gross income. Money received from the sale of capital assets used in the business generally is not included with receipts in the income statements since such income usually is not produced or earned during the period.

(3) **Expenses :** All expenses and costs involved in operation of the business during the period covered by the income statement should be included in the income statement. Thus all operating and fixed expenses are included. However, capital expenditures to purchase fixed and working costs such as real estate, machinery, milk cows, and breeding stock are excluded since such items casually are used in the business for several years. The depreciation on these items, which occurs during the period convered by income statement, is an expense, however, and should be included. Operating (variable) and fixed costs customarily are also shown separately in the income statement. Receipts are derived from sales of crops a lives stock and live stock products included figures are useful in analyzing the business.

(4) **Net income :** The net income (or loss) figures are useful in analysis of the business.

(a) **Net cash income :** Equals the cash receipts less cash expenses during the period covered by the statement. This is useful in analysis of cash flow and in preparation of income tax returns of the business.

(b) **Net operating income :** Is computed by subtracting operating expenses from gross profit. This measure of income facilitates comparison of farm with different fixed cost structures, i.e., amount of mortgage debt, different depreciation schedules and like. It also facilitates comparing income from operations on the same farm over a period of years.

(c) **Net farm income :** Is computed by deducting fixed costs from net operating income. It represents the income occurring to the operator and family labour management and equity capital of the three measures. Net farm income is perhaps the most useful. Net farm income represents amount available for firmly living income taxes and savings.

Thus income statement is basically comprised of three parts.

1. Receipts 2. Expenses 3. Net income

Farm Business A : Income Statement for the year 2005

		(Rs.)
(I)	Cash receipts	
	Livestock, sales of crops, govt. payments	56000
	and miscellaneous income	
*	Gross cash receipts	56000

*	Increase (Decrease) in current inventory	21000
*	Gross income	77000
(II) Less : livestock and livestock feed purchased		
*	Gross profit	55000
*	Total operating expenses	11000
*	Net operating income	44000
*	Total fixed expenses	15000
(III) Net farm income		29000

(5) **Financial Tests (Ratios) :** Four basic ratios which gives pertinent information on performance of the business are

    (i)   Capital turn over ratio

    (ii)   The gross ratio

    (iii) The operating ratio, and

    (iv) The rate of return on capital

(i) **The capital turnover ratio :** It is computed by dividing gross revenue by capital

$$\text{Capital turnover ratio} = \frac{\text{Gross revenue}}{\text{Capital}}$$

The capital turnover ratio indicates how much gross revenue has been received for each rupee of capital in the business.

(ii) **Gross ratio :** It is an overall measure of the income producing ability of the farm business. It is computed by dividing total expenses by gross revenue.

$$\text{Gross ratio} = \frac{\text{Total expenses}}{\text{Gross revenue}}$$

This ratio has the advantage of comparability from one farm to another and from year to year for the same farm.

(iii) **Operating ratio :** This ratio pertains to the current period of operation. It is obtained by dividing total operating expenses by gross revenue.

$$\text{Operating ratio} = \frac{\text{Total operating expenses}}{\text{Gross revenue}}$$

(iv) **Fixed ratio :** It is obtained by deviding fixed expenses by gross profit.

$$\text{Fixed ratio} = \frac{\text{Fixed ratio}}{\text{Gross profit}}$$

The operating ratio and fixed ratio combined together comprised the gross ratio.

The gross ratio, including both operating and fixed costs, pertains to the long run situation where as the operating ratio pertains to short-run situation. The fixed ratio is measure of the importance.

(v) **Rate of return on capital :** The rate of return on capital is obtained by dividing net income by the amount of capital

$$\text{Rate of return on capital} = \frac{\text{Net income}}{\text{Capital}}$$

* *Compounding :* It is the process by which the future value of the present income can be determined.

* *Discounting :* It is the process by which the present value of the future income can be determined.

PRICING POLICY

Introduction

According to the Law of Economics, it is stated that price, demand and supply are interdependent variables. When demand is more than supply, the price tends to rise in the market. When demand is less than supply, there is a tendency for the price to decrease. The change in demand and supply are reflected in the change in price in the market, according to the law of supply in Economics. Similarly when price increases, demand falls and when price decreases, demand increases.

The modern authors and the Economists have shown that the above relationship does not hold good always. There are many cases when change in the demand or supply is not the cause of change in price. We give below some of the instance of this type :

1. When income increases, people shift spending from food, fruits and textiles to canned food and durable like refrigerators, gas stoves, radios etc. So even if prices are constant, demand for canned food increases and demand for ordinary foodgrains decreases.

2. According to the Laws of Economics, when price increases demand is expected to fall. But in case of articles, which have sow value appeal, higher price increases the demand for the product. When price increases demand also increases for such goods. At least demand is constant in spite of increase in price, e.g., jewellery, refrigerator, gold, ornaments etc.

3. In case of certain goods, increase in price is followed by increase in demand. Sometimes higher price increases customer's confidence in quality. Customers expect that with increase in price, the quality has also been improved by the manufacturer. So, increase in price is followed by increase in demand.

4. Monopolists hold up prices even if demand goes down. They have no price competition. Here also law of demand and supply does not hold good.

Pricing is not so simple as explained by the economists. It is a very complex process. Normally, companies decide prices by taking into account various factors.

Situations when a Price is Set

A company is required to set prices in following situation :

(a) While introducing new product.

(b) While introducing established products in a new territory.

(c) When competitor changes prices.

(d) When demand changes or costs changes.

(e) When company has inter-related products that have inter-related demand or inter-related costs.

Methods of Pricing

In reality, there are four methods of deciding the price. These methods are as under :

(i)   Cost oriented price setting.

(ii)   Demand oriented price setting.

(iii) Competition oriented price setting.

(iv)   Product line pricing.

Let us describe these four methods in detail.

(i) Cost Oriented Price Setting

A. **Mark up Pricing :** Price is determined by adding some fixed percentage to the unit cost - say 10% or 15%. The manufacturers find out the cost of production, and add some percentage say 10% to that cost of production and arrive the price to be charged to the customers. The trader finds out the cost of purchase and adds some percentage say 10% to the cost of purchase and arrive at the price at which the goods are to be sold.

Retail pricing, groceries, furniture, clothing, jewellery, construction, military weapon, frozen floods, coffee, canned foods etc., average from 10 to 15%. Mark up (profit) should vary inversely with costs and with turnover prices are lower on re-seller's brands. Prices are however, higher on manufacturer's brand.

Benefits of the Cost Oriented Pricing

(a) In cost oriented pricing there is less uncertainty for the seller and the buyer. Costs can be easily determined and so it is a simple method of price setting.

(b) Prices are similar when costs are similar in the particular industry. There is less ambiguity about prices when costs can be compared.

(c) It is considered to be a fair method of determining the price. Cost oriented price setting is a method, which is fair to the buyer as well as to the seller. There is no exploitation of anybody when prices are changed according to this method.

B. **Target pricing :** Target pricing is second type of cost oriented pricing. The firm tries to determine the price that would give it a specified targets rate of return of its total costs at an estimated standard volume. For example, suppose the manufacturer has invested 2 crores of rupees, he wants 15% rate of net return. He finds out that amount to return on investment and determining the prices accordingly. Similarly, trader finds out his investment. Suppose he expects 15% net rate of return, he finds out the amount of return and determines the prices accordingly. Thus, in this method we determine the return first and than determine the prices later. It may be mentioned that long range average rate of return is about 15% to 20%. All public utility concerns are monopolists and they decide the prices by target pricing.

Benefits of Target Pricing : This method has three benefits —

(a) This method is fair to buyers. Buyers are not exploited in this methods.

(b) The method is fair to the investors because inventors get a fair rate of return on their investments.

(c) The method is very simple. The process of price setting has been simplified in this method.

This method has one important defect. The sellers not take into account the important factor that price do affect sales volume, when they decide the prices by this method.

Public utility concerns decide prices by this method. Electricity Board, Water Dept., Gas supply, Petroleum products etc.

(II) DEMAND ORIENTED PRICING

Demand oriented approach takes into account the intensity of demand. When demand has more intensity, the seller charges higher prices. When demand is price sensitive, the seller charges, lower price and attracts more demand. This method is also called price discrimination.

There are five types of price discrimination which can be attempted by the sellers:-

(a) **According to customers :** When price discrimination is according to customers, different prices are charged to different customers. Doctors charges different prices to the rich and the poor patients. Normally they charge higher price to the rich because their sensitivity to price is less. Since the sensitivity of poor people for price is more, the doctors charge lesser price to the poor customers. This pricing method is followed by pleaders, professionals and tax consultants.

(b) **According to product :** Price discrimination can be done according to product version also. Different versions of the product are manufactured with slight variation. For each variation separate price is charged. For example, refrigerator with top has a different price. The price difference cannot be explained by the increased costs. The manufactures of toilet soaps make different types of soaps and charge different prices according to degree of sensitivity of demand. The difference in price cannot be explained by difference in costs. The sensitivity of the customers for price is the basis for price discrimination according to product variation.

(c) **According to place :** Price discrimination can be attempted according to place. In theatre, normally there are three classes - Balcony, Upper stall and Lower stall. The tickets are different even if cost for screening the film is the same in each class. Similarly, one company can charge different price in different regions according to the sensitivity of the customers in different regions. For example, in a region of rich customers, higher price can be charged. But the same product is sold at lower price in a region, which mainly consists of poor customers.

(d) **According to time :** Price discrimination can be attempted on the basis of time. The hotels in a hill station charge higher prices in the summer season, but lower prices are charges in the off season. The electricity company in America charge different prices on different days and different times of the day. Similarly, in some countries public transportation has higher ticket in peak hours and lower ticket in lean hours. In India, telephone department charges higher prices during the day and lower price during the night for the same trunk call.

(e) **According to uses :** Price discrimination can be done according to uses also. Even if the product or the service is the same, the seller charges different prices for different uses. Electricity has lower rate for domestic use. Electricity has different rate for factories and for domestic houses.

There are certain conditions which must be fulfilled if price discrimination is to be done successfully.

(a) **Market should be segmentable.** If market cannot be segmented price discrimination is impossible.

(b) It should not be possible for the customers to resell the product. Otherwise, customers in the lower price segment would resell the product to the customers in the higher price segment and would make profits. In such cases price discrimination should be a failure.

(c) Competitor should not under-sell in the segments where the seller is charging higher prices for the products. Otherwise, price discrimination will not be possible.

(d) The seller has to keep different segments of the market separate, if he is to charge different prices in different segments. He is to ensure that the customers do not transfer the goods from lower price segments to higher price segments. Cost of segmentation and pricing should not exceed the extra revenue which the seller gets on account of price discrimination done by him in different segments of the market.

(e) Price discrimination should not affect the long term relationship between the seller and the customers. Customer satisfaction is the most important thing. It is the focal point. If customers resist, then price discrimination is nor possible.

(III) COMPETITION ORIENTED PRICING

The firm tries charges price as per the competitor's price - lower or higher. Even if costs changes, we maintain prices if competitors also maintain prices. But we change the price when competitor changes the price even if our costs have not changed.

There are two types of competition oriented prices :

(a) **Limitative Pricing :** The firm tries to keep its price at the average level charged by the industry. For homogenous products, there is no possibility of price discrimination, e.g., steel of TISCO, Mysore Steel or Bokaro Steel. So industry takes collective action to raise price or to lower price. One manufacturer charges price and all other competitors also change price accordingly. One company leads others follow. In paper industry also prices are changed by this method.

(b) **Sealed Bid Pricing :** Firms compete for jobs on the basis of bids (for example, Govt. jobs), manufacturer gets tenders for spare parts. To get a contract, firms try to give lower price than competitors. To raise prices means increase in profit, but it reduces the possibilities of getting contracts. So your skill lies in guessing the probability of getting contract at various bidding level and the skill in getting information about competitors' bids and in keeping our bid a secret.

There are three ways of deciding the price :

(1) Past bidding history

(2)   Trade gossip

(3)   Conjecture

If reliable data is received we can formulate model of probability.

(IV) PRODUCT LINE PRICING

When one product is a member of group of products, the pricing is very complicated. There are two cases of this type -Inter-related demand and Inter-related costs. Let us see how price is charged in case of products which have inter-related demand and which have inter-related costs.

1.   **Inter-related demand :** When the price of one product affects the demand for the other product, then we say that these products have inter-related demand. For example, demand for colour T.V. and black and white T.V. When price of a colour T.V. is lower, demand for black and while T.V. sets is affected. Demand for T.V. also affects the demand for components for colour sets and black and white sets. When price is lower for colour T.V. sets, demand would go up.

Soaps of different types are in real sense substitutes. Similarly, tooth pasts of different types also can be described as products having inter-relate demand. Higher prices can be charged for better styling, better qualities and extra lectures. Sometimes a super prestige model of a version of a product is introduced by the manufacturer to promote the other products in the lines of products. We charge different prices for different versions of the product to achieve the greater overall revenue for the manufacturer.

2.   **Inter-related cost :** Two products are called inter-related products in cost when a change in the production of one affects the cost of the other. Normally, by-products or joint-products have inter-related costs. A products using some production facilities are inter-related on the cost side even if they are not joint-products. In such cases accounting practices requires full allocation of costs. For example, sugar, molasses (alcohol), paper (bagasse) and bricks from dust are joint products with inter-related costs.

When we have inter-related demand to inter-related costs, in the line, we can determine the price of different products in the line through one of the following methods.

(a)   First method of pricing is to find out the full costs of different products and then decide the prices which are proportionate to the full costs of different products in that line. This method is followed by some companies. For example, a group of factories manufacturing sugar, paper, alcohol, furfural and bricks find out the costs of these different products and charge prices which are proportionate to the cost of each product in the line.

(b) The second method is to find out the incremental costs. Incremental cost is the addition to the total cost on account of increase of one unit of production. We manufacture 1000 tonnes of sugar and the cost of production of sugar is Rs.400 per quintal. Then we decide to increase the production of sugar by 1 tonne. We find out how much increase in cost occurs due to increase in 1 tonne of sugar. This is called incremental cost. This incremental cost is also found in case of other products in the line. For example, alcohol, paper furfural and bricks. The second method of pricing is to set prices that are proportional to incremental costs.

(c) The third method of pricing is to set prices that are proportional to conversion costs. Conversion costs are defined as the costs required to convert the raw material into finished products. We find out the cost of converting sugarcane into sugar. Then we find out the cost of converting molasses into alcohol. We also find out the cost of covering bagasse into paper and then we decide the prices of these products accordingly.

4 **Pricing Objectives** : A company must be clear on what it is trying to achieve in the way of overall business and marketing objectives before it can set up the price of the product. There are three pricing objectives— (a) Profits, (b) Sales Revenue and (c) Market Share.

Each possible price of a particular product has a different implication for profits which the company would earn by selling the products at that price, sales revenue the company would earn by selling the product at that production and the market share which the company would record in the market.

(i) *Profit Maximising Pricing* : One of the most common pricing objectives is to maximize current profits. Economists have worked out a simple model for pricing in order to maximize current profits, which the company would enjoy by selling the product at a particular price. This model can be used if company can acquire knowledge of its demand and costs of the products in question —

$$Z = R - C$$

Here Z is used to show the total profits. The total profits are the different between total revenue and the cost. R is the total revenue which is equal to price multiplied by the quality of the product sold by the company wants to sell. But it has some limitations in practice.

(a) This model assumes that the other marketing-mix variable are constant. In practice other marketing variable may not be constant.

(b)   This model assumes that competitors do not change their prices. In practice when we change our prices, our competitors also are likely to change the prices.

(c)   This model ignores the reaction of other parties in the marketing system. For example, government, suppliers, dealers etc. Actually they all react to various prices that might be charged. We cannot ignore their reaction.

(d)   This model assumes that the demand and the cost function can be properly estimated. In fact, it is extremely difficult to estimate the demand in advance or the cost in advance. So this model is not s useful. Besides, nowadays, companies emphasise on getting reasonable profits and not maximum profits.

(ii)   ***Market-Share Pricing*** : A company should select price that would maximize its market share. Sometimes in order to maximize market share, the company may forgo current profits. An increasing number of companies believe that long-run profitability of the company increases when market share of the company also increases. When market share of the company also increases. So, they are prepared to lose money for the first few years, but make up later when they dominate the market.

(iii)   ***Sales Revenue*** : Some companies believe that they should increase the sales revenue as much as possible. They fix up the price in such a way that the sales revenue can be maximized.

Market penetration pricing means to charge low price in the beginning and establish the product in the market. When the product is established the price is increased. Sometimes companies lose money in the beginning with penetration pricing. But they believe that they can make up that money later when they dominate the market, a low price will stimulate a more rapid market growth. This pricing sensitive market, a low price will stimulate a more rapid market growth. This pricing policy is also effective when the unit price of production and distribution fall with accumulated production experience. A low price would discharge actual and potential competition.

Market skimming is a policy when a firm charges very high price in the beginning as the product has high pricing value to the customers. So customers are initially prepared to buy the higher priced product. When there are enough buyers the demand is relatively inelastic and the market skimming policy should be followed. The unit production and distribution costs of producing a smaller volume are not so much higher that they cancel the advantages of charging higher price. When there is little danger that the high price will encourage new competitors in the market, this policy can be followed. Sometimes customers are under the impression that high prices means a superior product. In such cases also market-skimming policy can be followed.

As time passes, the firm will lower its prices and attract more price-elastic customers. This pricing policy has been followed for calculators, computers etc.

MARKETING MANAGEMENT

Meaning of Marketing

(i) Marketing is a place as an open space of a large building, where actual buying and selling takes place.

(ii) Marketing is the act of buying and selling.

(iii) An assembly of meeting together of people for their private purchases and sells of goods at a particular time and place.

Definitions of marketing

1. Marketing includes all activities involved in the creation of place, time and possession of tilities; place utility is created, when goods and services are available at the places, they are needed. Time utility when they are needed and possession utility, when they are transferred to those who need them.

2. Marketing include those business activities involved in the flow of goods and services from production to consumption.

Importance of Agricultural Marketing

1. Agricultural marketing plays an important role not only in stimulating production and consumption but in accelerating the place of economic development

2. Technological breakthrough has led substantial increase in agril. production, which result into larger marketable and marketed surplus. To main this tempo and pace of increased production through technological development, an assurance of remunerative prices to the farmer is a prerequisite and this assurance can be given to the farmer by developing an efficient marketing system.

3. The development of an efficient marketing system is important in ensuring that scarce and essential commodities reach different classes of consumers.

4. Marketing is not only an economic link between the producers and consumers; it maintains a balance between demand and supply.

5. The objectives of price stability, rapid economic growth and equitable distribution of goods and services cannot be achieved without the support of an efficient marketing system.

Importance of Agricultural Marketing in Economic Development is as follows

1. **Optimization of Resource use and output management :** An efficient agril. marketing system leads to the optimization of resource use and output management. A well-designed system of marketing can effectively distribute the available stock of modern inputs and thereby sustain a faster rate of growth in the agricultural sector.

2. **Increase in farm income :** An efficient marketing system ensures higher levels of income for the farmers by reducing the number of middlemen of by restricting the commission on marketing services and the malpractice's adopted by them in the marketing of farm products. An efficient system guarantees the farmers better prices for farm products and induces them to invest their surplus in the purchase of modern inputs so that productivity and production may increase.

3. **Widening of markets :** A well-marketing system widens the market for the products by taking them to remote corners both within and outside the country, i.e., to areas far away from the production points. The widening of market helps in increasing the demand on a continuous basis, and thereby guarantees a higher income to the producer.

4. **Growth of Agro-based industries :** An improved and efficient system of agricultural marketing helps in the growth of agro-based industries and stimulates the overall development process of the economy. Many industries depend on agriculture for the supply of raw materials.

5. **Price signals :** An efficient marketing system helps the farmers in planning their production in accordance with the needs of the economy. This work is carried out through price signals.

6. **Adoption and spread of New Technology :** The marketing system helps the farmers in the adoption of new scientific and technical knowledge. New technology requires higher investment and farmers would invest only if they are assured of market clearance.

7. **Employment :** The marketing system provides employment to millions of persons engaged in various activities, such as packaging, transportation, storage and processing. Persons like commission agents, brokers, traders, retailers, weigh men, hamals, packagers and regulating staff are directly employed in the marketing system. This apart, several others find employment in supplying goods and services required by the marketing system.

8. **Addition to National income :** Marketing activities add value to the product thereby increasing the nations gross national product and net national product.

9. **Better living :** The marketing system is essential for the success of the development programmes which are designed to up lift the population s a whole. Any plan of economic development that aims at diminishing the poverty of the agricultural population, reducing consumer food prices, earning more foreign exchange or eliminating economic waste has, therefore, to pay special attention to the development of an efficient marketing for food and agricultural products.

10. **Creation of utility :** Marketing is productive, and is as necessary as the farm production. It is, in fact, a part of production itself, for production is complete only when the product reaches a place in the form and the time required by the consumers. Marketing adds cost to the product, but, at the same time, it adds utilities to the product. The following four types of utilities of the product are created by marketing.

    (a) *Form utility:* The processing function adds form utility to the product by changing the raw material into finished form. With this change, the product becomes more useful.

    (b) *Place utility :* The transportation function adds place utility to the products by shifting them to a place of need from the place of plenty. Products command higher prices at the place of need than at the place of production.

    (c) *Time utility :* The storage function adds time utility to the products by making them available at the time when they are needed.

    (d) *Possession utility :* The marketing function of buying and selling helps in the transfer of ownership from one person to another. Products are transferred through marketing to persons having higher utility to persons having low utility.

Decisions in Marketing of Farm Products

Marketing management involves decision on four important areas, which are often referred to as marketing mix. They are decision on

1. Product
2. Price
3. Promotion, and
4. Place

Product Decisions

Product decisions are the most important ones as they are the heart of the marketing management.

The product decisions should have a strong market orientation. While developing a product the needs, the requirements of the consumers should be assigned top priority. Product decisions should also be based on the life cycle of the product. Sales and profits in a business often related to product life cycle. There are several distinct phases in the life of a product from its development and initiate introduction to its eventual removal from the market. They are

1. **The Development stage :** It is the period when market is analyzed and the product is developed.

2. **The introductory stage :** It is the period when the product appears in the market for the first time. Marketing strategy in terms of advertisements designed provide all information about the product is required. The new product associated with the high cost because of heavy promotional activities.

3. **The Growth stage :** It is a period of rapid expansion in sales and profits. the product will be available to larger sections of the society. Since, the fixed cost spread larger sales, profits will increase. Increased profits often attract new competitors. The presence of the competitors is felt in the market.

4. **The maturity stage :** This stage is characterized by slow growth in sales often even some decline in the sales as the market becomes saturated. Marketing strategy in terms of reducing price, refining the product, changing the product design adding new features to the product or increasing promotional efforts is required. As a result of these strategies cost may go up and reduce the profits.

5. **The declining stage :** This stage is characterized by rapid decline in sales and profits. Changing consumer preferences, availability of new substitutes may lead to death or removal of the product from the market. Some firms may withdraw from the market.

    The life cycle for each product may look very different from one product to another. Some products have very short life cycle lasting only a year or two while others may have life cycle that spread over a dozens of years. One job of marketing manager is to prolong the profitable life stage of the product.

Stages of Product Adoption

The manner in which customers adopt a new product is important for developing marketing strategies. It suggests how the new product should be introduced in the market. The stages of product adoption are:

1. **Awareness :** At this stage the customers have heard about the product but lack sufficient information to make purchasing decisions.

2. **Interest :** A potential customer starts to make comparison between the new product and old product.

3. **Evaluation :** Potential customer starts to make comparison between the new product and old product.

4. **Trial :** Customer samples the product.

5. **Adoption :** Customer integrates the product into a regular use.

The marketing strategy of agribusiness is to push the new product to adoption stage as quickly as is possible. In the first stages, more advertisements and promotional efforts are required, while in the next two stages distributing free samples, demonstrations are required.

Price Decisions

Pricing the product/service is a critical marketing decision because it influences the revenue generated. Some of the commonly used pricing methods are:

1. **Cost Pricing :** It is a simple method of adding margin to the use of a production of a product or service. The margin intends to cover overhead and handling cost and leave a profit.

2. **Market pricing :** This method simply sets price at the 'going rate'. In agri-business this method of pricing is widely used as the market is dominated by large competitors.

3. **Contribution to overhead pricing :** In this method the price of a product is fixed slightly more than the variable cost with a view to encourage additional sales over and above the normal sales.

4. **Penetration pricing :** In this method of pricing the product is offered at a low price in order to increase sales and wider acceptance quickly. This strategy is used at the time of introducing the new product into the market. After the new product has gained wider acceptance the price may be gradually raised.

5. **Skimming the market :** This is almost opposite to penetration pricing. In this method the product is introduced at a high price for affluent customers. The price is gradually lowered bringing the product into the range for less affluent customers. New superior hybrid varieties of seeds are after introduced using this strategy.

6. **Discount pricing :** In this method the product is sold at reduced price from the listed price Bulk discount pricing, cash discount pricing are also used in this method.

7. **Loss leader pricing :** In this method of one of the products is sold at a reduced price when product are purchased in combination. The idea is to encourage long run adoption of one of the products.

8. **Psychological pricing :** In this method psychological illusion about the price of a product is created. The prices set emotionally, satisfies the consumers.

9. **Prestige pricing :** Prestige pricing on the other hand appeals to a quality and elite image many people have tendency to equate the price with quality on an emotional basis. Some seed companies use this method of pricing.

Promotional Decisions

Promotional activities are designed to increase sales. These activities inform the customers about the various aspects of the product and convince them to buy the products. Advertising, personal selling efforts, general publicity programmes and sales support programmes are taken up in promotional activities. Promotional activities must be based on product life cycle, stage of adoption, competitor's action and availability of budget.

Place Decisions

Place divisions are concerned with the selection of appropriate methods and channels of distribution new work. Marketing channels are systematic ways of transferring product and ownership of the product from producer to the consumer. The cost incurred in each channel and the profit margins of the middlemen should be closely examined before selecting a particular channel. A firm may select a combination of channels starting from having its own retail outled to a involvement of wide range of middlemen.

MARKETING COST AND MARKETING EFFICIENCY

Marketing cost

Marketing cost comprises all the expenses incurred by several agencies engaged in marketing (functionaries) plus their margins right from the time the produce leaves the producer to the time the delivery is made to consumers.

These expenses are mainly handling charges at assembling, wholesaling and transshipment points; cost of storage and transport as required in the chain, cost of grading, packing and financing marketing activities and similar other items. They also include the profit margins of the intermediaries who help in bringing the produce from producer to consumer.

Cost of Marketing of Farm Products

The cost of marketing in case of farm products like grain, cotton, sugarcane etc. begins, when they leave the field or in some cases, when they are prepared for marketing on the farm. When marketing of agril. products takes place at the farm, the price paid to the farmer is generally considered as the cost of production and the marketing expenses are considered as the cost of production and the marketing expenses are considered to begin at this point. It is not possible to measure accurately the cost of marketing, when it takes place at the cost of marketing, when it takes place at the farm and as they are very small in amount, the price paid to farmer for his product either at the farm or at local shipping point is generally considered as the cost of production and marketing expenses are considered to gain at this point.

Factors Affecting Cost of Marketing

Factors affecting cost of marketing are many and vary from time to time, between different markets and with different products. The important factors are given below.

1. **Perishability:** More the perishability, more are the marketing costs, i.e., positive relationship.

2. **Extent of loss in storage and transportation:** More the losses, more is the cost.

3. **Volume of product handled :** More the volume handled, and then per unit cost is reduced, i.e., inverse relationship.

4. **Regularity in the supply of the product:** Cost is less, when supply is regular as compared to that when, supply is irregular.

5. **Extent of packaging:** Better and more packaging add to the cost-positive relationship.

6. **Extent of adoption of grading:** Cost of ungraded product is more as compared to graded one as ungraded product may necessitate more market services to be rendered.

7. **Necessity of demand creation:** Positive relation with cost.

8. **Bulkiness of the product:** Positive relationship.

9. **Need for retailing:** Positive relationship.

10. **Necessity of storage:** With storage cost increases.

11. **Extent of risk :** More the possibility of risk more is the cost, i.e., positive relationship.

12. **Facilities extended by dealers to the consumers:** More the facility more is the cost.

Reasons for higher marketing costs of Agril. Commodities

Agricultural products present peculiar features as follows. These circumstances cause for higher marketing costs.

1. Agricultural production is scattered over wide areas (on large number as well as small farms) due to which assembling of small lots into sizeable ones becomes an additional functions.

2. Agricultural products vary in quality and therefore, need grading which is again difficult. Some agricultural products are more bulky and perishable and therefore, more difficult to transfer and store.

3. Due to seasonality in production, the supply is irregular and therefore, the products are required to be stored to meet the continuous demand. At the same time as these cannot be consumed in raw form may need processing.

4. Greater time lag between initiations of production, receipt of product and then consumption involves sinking of money and over time there is possibility of risk due to physical deterioration and price fluctuation. This makes functionaries to resort to larger margins.

5. In foodgrain marketing, there is in general large number of middlemen as there is no restriction for their entry into the trade as compared to industry.

METHODS OF MEASURING MARKETING COSTS AND MARGINS

1. Lagged Margin Method

By following a particular lot of a commodity from the time it leaves the producer to the time, when it is put in the hands of final consumer and apportioning the shares of producers and various intermediaries in the price paid by the ultimate user. The method is known as lagged margin method. This is the best method as it takes into account the time elapsed between sale and purchase amongst different parties (producer, merchants, consumers etc.). Lagged margin is the difference between price received by a seller at a particular stage of marketing and price paid by him at a preceding stage of marketing during an earlier period. (Length of time between two points denotes period for which seller held the commodity).

2. Concurrent Margin Method

It consists of taking difference between prices of a particular commodity in the primary, wholesale and retail markets (i.e., successive stages of marketing) at a given point of time and then apportioning the differences in prices amongst various intermediaries and thus, knowing marketing costs and margins (i.e., the difference between farmer's selling price and retail price on a specific date is the total concurrent margin). Concurrent margins do not take into account the time that elapses between purchase and sale of the product.

Marketing efficiency : Efficiency many be defined broadly as the effectiveness or competence with which a marketing structure performs its designed function.

Marketing efficiency can be defined as the maximization of consumer's satisfaction with the least costs incurred in providing that satisfaction through the system of marketing.

But there is no way to precisely measure the satisfaction and marketing costs, taken by themselves, whether sizeable or small, give no indication of marketing efficiency just as the proportion of the farmer's share in the consumer's rupee gives no indication whether marketing is efficient or not. However, efficiency in marketing operations can be brought about the by keeping the costs down, development of market structure, reduction in risks; and timely distribution of commodities

Type of Marketing Efficiency

Marketing efficiency can be conveniently divided into:

(1) Technical (operational) efficiency (2) Economic (pricing) efficiency.

1. **Technical efficiency :** Technical efficiency pertains to the improvement of marketing structure, the utilization of the best method available for every marketing job, and the use of these methods with maximum effectiveness.

   Use of trucks in place of bullock carts, of mechanical aid and balance for hand, of phone for messenger etc. are some of the examples of technical efficiency.

2. **Economic Efficiency : (Pricing) :** Economic efficiency pertains to the minimization of the amounts of inputs required for a given output of goods and services, involving elimination of waste, high costs and exploitative profits.

In a competitive market or conditions close to perfect competition, there are maximum possibilities of minimizing waste and exploitation. Such a condition leads to the fixation of a uniform price throughout the market area.

The measurement of marketing efficiency : It is possible by a functional analysis of the cost incurred on the marketing processes by all the middlemen involved, by all the producers and consumers of the products on the one hand, and by the value added to the products by the marketing organisation, on the other. This total cost divided by the total value of the products sold (value added by marketing) provides a percentage figure, which accurately measures market efficiency. It can, therefore, be computed as

$$\text{Marketing Efficiency} = -\frac{(\text{Value added by marketing organisation} \times 100}{\text{Cost of marketing services}}$$

If in a market, the total marketing cost incurred by all participants (producer, middlemen and consumers) were Rs. 6000 and the value added by marketing were Rs.9000 the efficiency of market could be measured as Efficiency = 9000 × 100 = 150/6000.

MARKETING COST AND MARKETING EFFICIENCY

Generally, high marketing costs and margins are considered to be indicators of inefficiency in the marketing process. But this not always true. The fact that a major part of consumer's rupee is spent on marketing costs does not always mean that something is wrong with the distribution system. A number of factors may operate to cause high proportion of marketing costs, without any reflection on the efficiency of the marketing system.

1. **Place of production :** The geographical localization of production brings about a change in the marketing cost. Higher marketing cost in area away from the market is not a reflection of inefficiency of marketing system.

2. **Time of production :** Food articles are made available throughout the year, which is possible only by storage. This adds to the cost. The cost of goods sold in the off-season is higher than that in the peak season.

   Increase in the marketing cost in lean season than the post harvest season of commodities does not reveal decrease in the marketing efficiency.

3. **Form of product :** The cost of marketing of a processed products is higher than that of a raw product. But the high marketing cost because of processing is not an indicator of the inefficiency of the system.

Hence, the levels of marketing costs and margins of the share of the producer in the price paid by the consumer should be used as measures of marketing efficiency with some care.

MARKET SEGMENTATION

1. The Nature of market

The heart of any consumer market is people. Similarly, the foundation of an industrial market is the companies that use its products. While people and companies determine the breadth of a given market, they are not the only factors involved. Purchasing power - that is, cash or the ability to borrow - is also a necessary condition for a market to exist. Many people would like a new Rolls Royce. But few have the financial capacity to buy one.

Another necessary condition is the desire or willingness to buy. There are a multitude of items that you could buy but don't. These are perfumes, toothpastes, shirts, and so forth that you simply don't like. You are not part of the market for those goods. Still another necessary condition for a market to exist is the authority to buy. Individuals is an industrial environment may have a strong desire for certain new equipment, but they do not have the authority to make the purchase decision. In short, a market is made up of people or companies with the ability and willingness to buy.

The scope of a market depends on many other factors as well. Substitute goods, or the lack of them, will influence a firm's share of a market. Generally speaking, the more substitutes there are for a company's product and the cheaper they are, the smaller the company's market share will be. An item that is unique or can do a job much more efficiently than older products will increase its market share. Complementary goods, such as shoes and socks, beer and pretzels, automobiles and tyres, will experience an increase in market share if demand for the complementary product increases. If the demand for tape recorder rises, so should its complement - cassettes.

In the industrial goods market the demand for machinery depends on the machine's output capability relative to its cost. The greater its equipment's productive capacity compared to its cost, the larger the industrial market. Again, other factors enter the picture, such as demand for the final product, the substitutability of labour or capital, and the number and quality of alternative machines.

2. Market Segmentation

The market for virtually any product is not an amorphous mass of people or firms but a set of subgroups with different needs and desires. Market segmentation is the process of identifying and evaluating various strata or layers of a market.

Market segmentation is the division of a market into these subgroups, which have special, needs and preference and which represent sufficient pockets of demand of justify separate marketing strategies.

It is difficult to think of any consumer or industrial product or service with a single, homogenous market. Knowledge of market segments relevant to the company's offerings is essential to the development of effective marketing strategies. Yet few companies attempt to serve all market segments. Even IBM, the dominant company in the electronic data processing industry, concentrated primarily on the major business and government segments, leaving to others (at least until now) the market segments for large engineering and scientific application, small business systems, and consumers uses. Going further, IBM segmented the business market by the type of industry such as manufactures, airlines, banks and news paper publishers. While each of these industries has common computer application needs (e.g., for accounting data and pay roll preparation), they also have different needs - manufacturers for inventory control, airlines for ticket reservations, banks for account retrieval, and newspapers for typesetting and makeup.

3. Evolution of the Market Segementation Concept

Companies have long recognized and responded to certain types of market segments such as price, quality, age, sex, geography, application and use. An early example was the organization of General Motors into the Chevrolet, Pontiac, Buick, Oldsmobile and Cadillac divisions, arranged to cover automobile price and quality segments from economy to luxury. This was in contrast to Henry Ford's approach. Ford has offered one standardized, low-priced model designed for the mass of potential car buyers.

Widespread acceptance of the concept of market segmentation occurred as the result of (i) increasing awareness of the almost infinite number of ways in which markets can be segmented and (ii) growing recognition that separate strategies for separate segments offer opportunities for increased sales and profits. The term 'market segmentation' has come to encompass not only identification and measurement of segments, but also the development of separate marketing strategies for dealing with these segments.

Consumer market segmentation is a natural response to the growth of discretionary income and population. Discretionary income enables people to include their preferences while large populations create groups of buyers with similar preferences of sufficient size to enable companies to serve them at a profit. Industrial market segmentation is a natural response to the growth of manufacturing and service industries, and government and private institutions required to serve a growing, affluent population.

4. Strategic Implications

Wendell Smith, in a pioneering article, introduced strategic implications of market segmentation by pointing out that companies have two basic options (i) to capture as large a share of the heterogeneous market as possible with a single product offering by attempting to differentiate its brand from all others in the market or (ii) to serve one or more of the homogenous segments with product designed to appeal to the preferences of each segment.

Combining Differentiation with Segmentation : While the strategic options pointed out by Smith do exist, most companies today practice market segmentation and at the same time, attempt to differentiate their brands from those of competitors within the segments served.

The soft drink is a case in point. The two leading brands - Gold Spot and Thumps Up - compete for the same market segments. Both companies have basically the same users (youth), flavour, low-calorie alternatives, packaging (individual and family sizes, cans and bottles), place of purchase (convenient goods outlets, vending machines, and soda fountains), and geographic locations (including foreign countries). Yet both companies also try to differentiate their brands.

5. Alternative Segmentation Approaches

There are three basic approaches a company can take to market segmentation :

(i)   serving primarily in large segments,

(ii) serving primarily the smaller segments, or

(iii) serving a cross-section of both

Concentration of Large Segments : Most product markets have a primary or "regular" market segment which accounts for a sizeable portion of total industry volume.

The advantage of concentrating on the larger or largest market segment is that it offers the greater sales potential. The disadvantages is that other major brands are also attracted to the larger segment; consequently, competitive activity is intense, making it more difficult for each company to attain or retain market share.

Concentration of Secondary or Smaller Segments : Concentration on a smaller segment offers less potential sales volume, but usually there are fewer competitors fighting for share.

In the electronic computer industry few companies have competed successfully with IBM for the general-purpose "mainframe" (central), systems business, which continues to represent the largest rupee market segment. Most of the companies that succeeded in gaining a foothold in the computer market did so by concentrating on smaller segments.

Serving Multi-segments

The prevailing pattern has been for larger firms to concentrate on major market segments and smaller firms to concentrate on lesser ones. Now larger firms are expanding into smaller segments as well. Multi-segment strategies are common also in the markets where the same product is sold to all segments.

6. Segmentation Criteria

Because no two buyers have precisely the same preferences, the smallest market segments would be the individual, family, or industrial buyer. From a practical standpoint, therefore, when we speak of market segments we are referring to groups of buyers with similar preferences. While there are exceptions—such as custom-built homes and machinery designed to the industrial buyer's specifications - a viable market segment in most cases must :

1. Be identifiable and describable in terms of the segment's needs and preferences.

2. Be measurable, so that the segment's size and demand potential can be determined.

3. Be large enough to enable both producers and distribution channel intermediaries to serve it profitably with a separate marketing strategy.

Some maintain that a segment should also be "reachable" with a marketing programme. While patently true, normally the segment can be reached if the above three criteria are met.

(a) **Identifying Segments :** Existing market segments can be recognized either by direct observation or by formal market research. They can also be created by developing new products that fulfill latest needs.

Most developed industries are already serving an already established segment is to complete for this segment. A second option is to develop a new product to create demand in a latent, previously unserved segment. We continue to see new market segments created by managements willing to engage in market research, creative thinking and experimental product and market development.

(b) **Measuring Segments :** Measurement of already established segments is relatively simple since it merely involves measuring existing sales patterns. Published industry data may be available or, if not, measurement can be obtained using standard market research survey techniques. Segment data covering some major consumer products can be purchased from private research firms. Measuring the sales potential of previously unserved or unrecognized segments is quite another matter, and usually requires sophisticated and costly market research. Expensive and time-consuming test marketing also may be needed to verify the research findings. Measuring and developing new market segments is similar to the steps involved in new product development.

(c) **Segment Size and Profitability :** Serving two or more market segments may be more costly for a company than treating the market as a single entity. The latter approach offers the cost efficiencies associated with mass production and the use of a single promotional programme. Segmentation, on the other hand, incurs the added costs of multi-production lines and multi-promotional programmes. Selling costs also rise when distribution must be forced through wholesalers and retailers who normally resist providing warehouse and shelf space for additional product lines. Nevertheless, market segmentation may be the more profitable approach. Higher costs may be more than offset by higher prices, greater total unit volume, or the effects of incremental volume in absorbing fixed costs. The decision to adopt a market segmentation strategy should be preceded by careful profit analysis. Once a segmentation strategy has been adopted, special care must be exercised in going after additional segments. Markets can be segmented into smaller and smaller units until they reach the point of no profit return.

7. Types of Consumer Market Segments

A comprehensive list would contain thousands of market segments. Each product industry offers its own opportunities for market segmentation and the segments for one industry may be quite different from those for another.

Let's look briefly at some of the segments in table :

(a) **Geographic** - Many companies market locally or regionally; have, unserved geographic segments offer potential opportunities for expansion. National companies also may move foreign markets, Geographic variations in weather, topography, water, soil, etc., offer segmentation opportunities for many types of consumer goods (e.g., clothing, sports equipment and tyres).

(b) **Demographic** - Most consumer markets can be segments by one or more demographic variables. Food and drink consumers can be categorized by age, race, ethic background, religion, health and physical characteristics; buyers of housing can be segmented by personal status; and so on, ad infinitum.

(c) **Economic** - Income is a useful of segmenting markets for many products and services, particularly higher-priced items like automobiles, housing and some furnishing. Experienced markets however, do not consider income as a sole factor when segmenting markets. Rather, they consider it along with social, psychological and other factors that also influence purchase behavior. Demand results from a combination of ability to buy and willingness to buy. Consequently, income is only one element of the demand equation. It is usually fruitful to look beyond income (and other contributors to purchasing power, such as savings, assets, and credit availability) for those other demand factors, which determine willingness to buy.

(d) **Social and Psychological** - Martine A.U. found social class to be a better indicator than income for segmenting markets for savings and investment, travel and home furnishings. It can even be a key factor in consumers' retail store selections. Reference groups also significantly influence segmentation. Each generation of college students, for example, adopts a mode of dress distinctly its own, yet quickly adapts to the mode of dress of postgraduate work environment. Stages of the life cycle as well as lifestyles influence purchasing patterns - for example, singles versus couples with children. Personality differences account for the great variety in the demand for items reflecting self-expression, such as clothes, jewellry, or art objects.

Social and psychological segments are easier to identify than they are to measure. The measurement challenge has attracted a number of experimental researchers.

(e) **Preference and Use :** Variations in usage rates represent an important means of segmentation. For purpose of marketing planning, the problem is to determine the heavy user's characteristic so that marketing strategies can be devised which will reach the heavy users. Soap and detergent usage rates may correlate with family size. But one can wonder about household cleaning compounds. In a private study we found that the frequently of kitchen floor, washing ranged from daily for families that considered "cleanliness next to Godliness" to almost never by those who felt that "a little dirt never hurt anybody."

Use applications from the basis for market segmentation for a great many products.

Variations in demand for many products are based on variations in attitudes towards quality and performance, which are related, at least partly, to price. With hobby equipment like cameras and sports equipment like tennis rackets, the neophyte often begins with lower-priced, lower quality-performance equipment and moves up the scales as interest and skill increase. Non-correlation between price and income is frequently seen in such products. The enthusiast often manages to find the money to obtain the finest equipment, irrespective of income level.

One idea of segmentation is based on benefits sought.

8. Types of Industrial Market Segments

Industrial marketers have long been aware of market segmentation because many goods and services are sold to different industries, each with its own products and service needs. There are also relatively homogeneous markets, which can be served with single product lines such as typewriters, lubricating oils, and air transportation. Even homogeneous markets, however, can de segmented by volume and profit potential.

As with consumer product markets, industrial product markets can be segmented in many ways, although each company need only analyse the particular markets it chooses to serve.

9. Common Sense Approach to Sementation

Most of the literature dealing with segmentation research applies (i) to consumer markets and (ii) to companies serving large mass markets. For a large consumer goods company, competing for market share with large competitors, any information, which will give it, an advantage is worth considerable investment in consumer research. Not all market segmentation decisions must be based on sophisticated and expensive primary market research; however, many segmentation opportunities can be evaluated with available data. One does not need research to know that a company operating in one geographic segment may consider the question of expansion into other geographic areas.

General Motors needed only observation and available statistics to know that the small car segment had grown to nearly one-fourth of the total car market, and that it was being developed primarily by foreign companies. Hanes did not need sociological research to recognize that low-priced hosiery was beginning to be sold in supermarkets - a segment; Hanes did not serve at that time. A manufacturer of polyethylene film, serving the package manufacturing market, learned from its customers that the privately owned segment - where he company's sales were concentrated - was declining and that the large publicity owned segment - to which it did no sell - was growing.

Basically, we should apply whatever methods are required for he particular situation. Sometimes this will mean sophisticated research; at other times, only common sense, analysis of available data, or relatively simple market research if required. And despite the fact that most literature deals with consumer markets, remember that segmentation applies to industrial market as well.

MANAGEMENT OF INVENTORY AND DETERMINATION OF LEVEL OF INVENTORY

Investment in inventory constitutes one of the major investments in current assets. The various terms in which a manufacturing firm may carry inventory are :

1. **Raw material :** There represent inputs purchased and stored to be converted into finished products in future by making certain manufacturing process on the same.

2. **Work in progress :** There represent semi-manufactured products which need further processing before they can be treated as finished products.

3. **Finished goods :** These represent the finished products ready for sale in the market.

4. **Stores and supplies :** These represent that part of the inventory which does not become a part of final but are required for producing process

eg cotton waste, oil and lubricants, soaps brooms, light bulbs, etc. These are a very minor part of total inventory.

MOTIVES FOR HOLDING INVENTORY

A company may hold the inventory with the various motives as stated below

1. **Transaction motive :** The company may be required to hold the inventories in order to facilitate the smooth and uninterrupted production and sales operation. There may be a time lag between the demand and for the material and its supply. Hence, it is needed to hold the raw material inventory.

2. **Precautionary motive :** The company may like to hold the inventories to against the risk of unpredictable changes in demand and supply forces supply of raw material may get delayed due to the factors like strike, transport disruption, short-supply, lengthy processes involved in import of the raw material etc. Hence, the company should maintain sufficient level of inventories to take care of such situations. Similarly, the demand for finished goods may suddenly increase and if the company is unable to supply them, it may mean rain of the competitors. Hence, the company will like to maintain sufficient stock of finished goods.

3. **Speculative motive :** The company may like to purchase and stock the inventory in the quantity, which is more than needed for production and sales purposes. This may be with the intention to get the advantages in terms of quantity discounts connected with bulk purchasing or anticipated price rise etc.

OBJECTIVES OF INVENTORY MANAGEMENT

Usually, the company is faced with the following conflicting objectives in the area of inventory management :

(i) To carry maximum inventory in order to facilitate efficient and smooth production and sales operations.

(ii) To minimize investment in inventory to maximize profitability.

The over-investment or under investment in inventories is undesirable as both involve the consequences.

The over-investment involves the consequences like.

(i) Unnecessary blocking of fund in inventories and hence loss of profit.

(ii) Excessive storage and insurance cost.

(iii) **Risk of liquidity :** The inventories once purchased and stored and normally difficult to dispose off at the same value. In other word, the value of inventory reduces with the increasing holding period.

The under investment involved the consequences like :

(i) It sufficient stock of raw material and work in progress is not available, it may result into frequent interruptions in production.

(ii) If sufficient stock of finished goods is not available it may not be possible for the company to serve the customers properly they may shift to the competitors.

It can be said that the objective of inventory management is to minimize the investment in inventory without affecting the production or sales operations.

Techniques of Inventory Management

1. Economic order quantity

2. Fixation of inventory levels

3. Inventory turn over

4. ABC Analysis

5. Bills of materials

6. Perpetual inventory system.

Economic Order Quantity

It indicates that quantity which is fixed in such a way that the total variable cost of managing the inventory can be minimized. Such cost basically consists of two parts. First ordering cost (which in turn consists of the costs associated with the administrative efforts connected with preparation of purchase requisitions, purchase enquires, comparative statements and handling of more number of bills and receipts). Second carrying cost, i.e., the cost of carrying or holding the inventory (which in turn consists of the cost like good rent, handling and upkeep expenses, insurance, opportunity cost of capital blocked, i.e., interest etc.). There is a reverse relationship between these two types of costs, i.e., if the purchase quantity increases, ordering cost may get reduced but the carrying cost increases and vice versa.

A balance is to be track between these two factors and it is possible at Economic order quantity where the total variable cost of managing the inventory is minimum.

It is possible to fix the Economic order quantity with the help of mathematical formula.

Let Q be Economic order quantity

A be Annual Requirement of material in units

O be cost of placing an order (assumed to remain constant irrespective of size of order)

C be cost of carrying one unit per year

Now, if A is the annual requirement and Q is the size of one order, the number of orders will be A/Q and the total ordering cost will be : A/Q x O. Similarly, if the size of one order is Q and it is assumed that the inventory is reduce at a constant rate from order quantity to zero when it is repurchased, the average inventory will be and the cost of carrying one unit per year being c, the total caning cost will be Q/Z x C

Thus,

Total cost = Ordering cost + carrying cost

$$\frac{A}{Q} \text{ and } O + \frac{Q}{2} \text{ and } C$$

The intention is that the value of Q should be such that the total cost should be minimum. Hence, taking the first derivative of the equation with respect H Q and setting the result to zero.

$$Q = \frac{2 \times A \times O}{C}$$

Where,

Q = order quantity

A = annual requirement in units

O = cost of placing an order

C = cost of carrying one unit per year

Example : A company uses 200 units of a component very month and it buys them from outside supplier. The order placing and receiving cost is Rs. 100 and annual carrying cost is Rs. 12 calculate EOQ

Solution : $\text{EOQ} = \dfrac{2 \times A \times O}{C} = \dfrac{2 \times 2400 \times 100}{12} = \dfrac{2 \times 200 \times 100}{12} = 40000 = 200 \text{ units}$

In some cases, the carrying cost may be expressed as an annual percentage of the unit cost of purchases.

$$\text{EOQ} = \frac{2 \times A \times O}{C \times I}$$

Where, l = carrying cost expressed as a percentage of unit purchase price

Example : Work out EOQ and total cost with the following data.

Annual Demand : 5000 units ordering cost : Rs. 60 per order

Price per unit : Rs. 100

Inventory carrying cost : 15% on average inventory

Solution :

$$EOQ = \frac{2 \times 5000 \times 60}{15\% \text{ of } 100} = \frac{2 \times 5000 \times 60}{15} = 2 \times 5000 \times 4 = 4000 = 200 \text{ units}$$

the total cost of managing inventory ordering cost + carrying cost

Ordering cost = A/QxO = 5000/200x60 = 25x60 = 1500

Carrying cost   = Q/2xC= 200/2 x 15% of 100= 100 x 15= 1500

Total cost = 1500+15 = Rs. 3000.

Example : Estimate the total cost of mar inventory for kupil orders will purchases gooo spare parts for annual requirements, ordering month usage at a time. Each costs Rs. 20. The ordering cost per order is Rs.15 and the carrying charges are 15% of the average inventory per year.

Solution : Number of orders = $\dfrac{\text{Annual requirement}}{\text{Order size}}$

     Order size = $\dfrac{\text{Annual requirement}}{\text{Number of orders}} = \dfrac{9000}{12} = 750.$

    number of orders = 12

Ordering cost = Number of orders x cost of placing an order

      = 12 x 15 = 180

Now, carrying cost = order size /2 x cost p

x carrying cost in %

      = 750/2 x 15% of Rs. 20 = 375 × 3 = 1125

total cost = ordering cost + carrying cost

      = 180 + 1125 = Rs. 1305

(1) **Fixation of Inventory Levels :** Fixation of various inventory levels facilitates initiating of proper action in respect of the movement of various materials in time so that the various materials may be controlled in a proper way. However, following propositions should be remembered.

   (i) Only the fixation of inventory levels does not facilitate the inventory control. There has to be a constant watch on the actual stock level of various kinds of materials so that proper action can be taken in time.

   (ii) The various levels fixed are not fixed on a permanent basis and are subject to revision regularly.

The various levels which can be fixed are as below

1. **Maximum level :** It indicates the level above which the actual stocks should not exceed. If it exceeds, it may involve unnecessary blocking of funds in inventory while fixing this level following factors are considered.

   (i) Maximum usage (ii) Lead time (iii) Storage facilities available, cost of storage and insurance etc. (iv) Prices for the material, (v) Availability of funds (vi) Nature of material, e.g., if a certain type of material is subject to Government regulations in respect of import of goods etc., maximum level may be fixed at a higher level (vii) Economic order quantity.

2. **Minimum Level :** It indicates the level below which the actual stocks should not reduce. It is reduces, it may involve the risk of non-availability of material whenever it is required. While fixing this level following factors are considered.

   (i) Lead time (ii) Rate of consumption

3. **Re-order Level :** It indicates that level of material stock at which it is necessary to take the steps for procurement of further lots of material. This is the level following in between two existences of maximum level and minimum level and is fixed in such a way that the requirements of production are met properly till the new lot of material is received.

4. **Danger level :** This is the level fixed below minimum level. If the stock reaches this level, it indicates the need to take urgent action in respect of getting the supply.

Calculations of various levels

(1) **Re-order level :** Maximum lead time x maximum usage

(2) **Maximum level :** Re-order level + Re-order quantity (minimum usage x minimum lead time)

(3) **Minimum level :** Reorder level - (Normal usage x Normal lead time)

(4) **Average level :** Maximum level + minimum level/2

Example : Two components x and are used as follows :

   Normal usage - 50 units per week each

   Minimum usage - 20 units per week each

   Maximum usage - 75 units per week each

   Re-order quantity - x - 400 units

                     Y = 600 units

Re-order period - x = 4 to 6 weeks

Y = 2 to 4 weeks

Calculate for each component :

(a) Reorder level

(b) Minimum level

(c) Maximum level

(d) Average stock level

Solution :

1. **Reorder level :**

    Maximum Level Time x Maximum usage

    X = 6 weeks x 75 units = 450 units

    Y = 4 weeks x 75 units = 300 units

2. **Minimum Level**

    Re-order Level (Normal usage x Normal Lead Time)

    X = 450 units - (50 units x 5 weeks) = 450 (250) = 200 units

    Y = 300 units (50 units x 3 weeks) = 300 (150) = 150 units

3. **Maximum Level :**

    Reorder level + Re-order quantity

    (Minimum usage x minimum lead time)

    X = 450 units + 400 (20 units x 4 weeks) = 450 + 400 (80) = 850 + 80 units = 770 units

    Y = 300+600 (20 units x 2 weeks) = 900 (40) = 860 units

4. **Average stock level :** Maximum level + minimum level/2

    X = 770 + 200/200 = 970/2 = 485 units

    Y = 860+150/2 = 1010/2 = 505 units

AGRO-BASED INDUSTRY

Preparation of Agril. Business Project

An Agril-Business project is an investment activity in which financial resources are expended to create capital assets that produces benefits over an extended period of time. It is an activity for which money will be spent in expectation of return and which logically seems to tend itself to planning, financing and implementing as a unit.

Project is specific activity with specific starting point and specific ending point. It will have a well-defined sequence of investment and production activities and a specific group of benefits that we can identify. The project normally will have a specific geographic location or a rather clearly understood geographic area of concentration.

Steps involved in preparation/setting up an Agri-business. Project

1. Identification of project ideas

2. Market Analysis

3. Technical and organisational Analysis

4. Financial and Economic Analysis

5. Feasibility Report preparation

6. Agro. Industry finance

7. Governmental Aid

8. Monitoring and Evaluation

1. Identification of project ideas

The first step involved in setting up of any agro-based industry is the identification of project ideas which can be done by the entrepreneur, either on his own depending on the necessary skills required and experience in concerned area or through aid of specialized agencies.

(i) Ideas for different projects will depend on requirements and availability of capital for the project from different sources.

(ii) The entrepreneur's background often tends to indirectly influence the idea generation as the individual limits himself to fields in which he has some sort of technical knowledge or experience.

(iii) The availability of raw materials determines the project ideas, i.e., easy availability less transport cost

(iv) Assessment of the market structures will also affect the ideas for a particular project as in case of existing monopolies. It is generally difficult to set up and establish industries.

(v) Availability of skilled manpower plays a very importance role in identification of project opportunities, as it is the human resources, which make for success or failure of any enterprise.

(vi) Projects which involve the wide fluctuations in input prices or the prices of raw materials are normally thoroughly assessed before setting up of project in those area and normally depends on the asset base of the entrepreneur.

(vii) Projects which do not lead to degradation of environment and fit into the various criteria of the air pollution, water pollution, regulation etc. are given the first priority as compared to the environmentally hazardous projects.

(viii) Investment in industries subject to govt. control like the entrepreneurs very often avoid sugar, milk powder etc. as there is a great degree of uncertainty involved in such projects.

2. Market Analysis

The market analysis is mainly concerned with assessment of potential market, the available market and targeted market, types of distribution channel to be developed, to workout a proper price structure for the produce and to evaluate the growth prospects for the industry.

(i) **Consumer analysis :** The consumer analysis mainly consists of the identification of economic, socio-cultural and demographic characters of the potential consumers, identification of needs of consumers for making purchase, identification of consumers mode of making purchases of the produce so as to arrive at the distribution channels to be used and finally assessment of necessary market information and the research techniques required.

(ii) **Marketing plan :** The next step is to carry out an in depth estimation of competition in the field such as no. of competitors and the location of their enterprises, the potential market and share of competitors and the role of substitute products in market.

(iii) **Marketing plan :** Keeping in mind the prospects for the product and the level of competition, an effective marketing plan should be evolved comprising of the following :

  (a) *Product Development :* The first step in the marketing plan comprises of the development of a quality product which aims at the competitive differentiation of the products and arriving at an appropriate product mix.

  (b) *Pricing mechanism :* An appropriate strategy for the systematic pricing of the produce should be adopted keeping in mind the objectives whether it is market penetration or break even or the quick sales etc.

  (c) *Promotion strategy :* The promotion strategy to be adopted based on the personal sales, advertisement, media to be adopted, cost to be incurred on promotion etc. need to be formulated.

  (d) *Distribution systems :* The number of distributors at each level, the types of wholesalers and the retailer, the cost, quality and dependability of existing services and the facilities etc. should be evolved.

(e) ***Integration of the Different elements :*** An integration of the different marketing elements as per the marketing strategy should be compatible with the company's financial, organisational, production and procurement operations

(iv) **Demand forecasting :** The demand forecasting involves collection and analysis of data to understand future market potential and projected market share of the proposed unit enabling entrepreneur to arrive at the project profitability, financial and raw material requirements etc. Several methods are used to determine the sale forecast such as the opinion of experts, the opinion of sales force who are in actual contact with the consumers.

3. Technical and Organizational Analysis

The technical analysis deals with the factors such as whether the scale of production is optimal based on the break even analysis and whether there will be full capacity utilization, selection of a suitable production process through appropriate technology, selection of the appropriate equipment and the machines, making provision for the treatment of the effluents, proper location of the plant with relation to the availability of the raw materials and technical manpower, care should be taken that the technology used is the appropriate technology and that modifications can easily be made when required without rendering the technology obsolete.

The organisational analysis is divided into preparation phases and operation phase. The preparation phase consists of all aspects including the installation of machinery, construction of building, commissioning of plant etc. The operation phase relates to plan of recruitment and training technical manpower, and procurement of raw material.

4. Financial and Economic Analysis

The financial analysis determines whether the project satisfies the investment criteria as the whether it shows an appropriate level of commercial profitability from the point of view of the entrepreneur.

The economic analysis of the project normally focuses on how to identify the effects of project on various segments of society and quantification of effects on various segments to reflect their values to the society.

5. Feasibility Report Preparation

The feasibility report of an investment proposal provides information required by Decision maker for appraising proposal, which is submitted to financial institutions for assistance and relevant govt. depts for necessary permission. Feasibility report contains general information of industry.

6. Financing of Agro-Industrial Project

Long-term sources of finance are for tangible fixed assets like the land and site development, building and civil works, plant and machinery etc. and for intangible fixed assets like preoperative expenses. Short-term sources of finance are available for inventories, receivables etc The institutions like Industrial finance corporation of India. LIC, NCDC, etc. are providing finance for setting up agro industrial projects.

7. Government Aid

The government is providing incentives to setting up of agro industries like financial grants and several subsides. Income tax relief's, facilities for import of raw materials. Entrepreneur can make use of the several facilities being provided.

8. Monitoring and Evaluation

This has been given in detail under appraisal/evaluation of Project.

ECONOMIC EVALUATION OF AGRIL. PROJECTS (Appraisal of Agril. Projects)

Post completion audit required projects are to be :

1. Identified—At the beginning
2. Formulation and then
3. Evaluation. Before taking investment project must be economically viable and technically feasible.

Economic Evaluation

The project is implemented when it is economically viable and technically feasible. It is concerned with sophisticated benefit-cost analysis.

Every project has go to the three phases.

(i) Identification and selection of the project

(ii) Formulation of the project

(iii) Appraisal/evaluation of project

Each phase involves several steps, (activities) analysis in depth and procedure conducted at various degree or depth or precision and findings are used to meet the requirements of the subsequent phase until the project is finally evolved.

Why there is need of project : Resources are scare why implement the project? Speedy growth project is part of the programme various projects involved in programme.

Because of the scarce resources and in order to speedy economic growth scare resources are optimally used. Economic and social development involving many activities in a wider strength. The choice of the project is strategic important in that selection in good the momentum can increase rapidly. The development process will be low. If the selection of the project will be poor and the additional development efforts will be handicapped precondition for development is efficient utilization of scarce resources.

Three M'S are important in project identification, formulation and evaluation of project

   (i)   Money

   (ii) Material

   (iii) Man power

   Sound project analysis called for three phases.

   (i)   A good identification

   (ii) Sound formulation of project

   (iii) Sound appraisal of project.

   1. **A good identification :** (i) Assurance that the project related would serve the needs of economy, which are of priority (ii) It is consistent with development targets of the sectors concerned or relevant. It ranks high among the competitive alternatives.

   2. **Sound project formulation :** It entails no of stipulations, it makes certain lag the (i) Objectives and targets of the projects are clearly defined. (ii) Cost and benefits are properly identified and competent (iii) Supply of needed input is assured that the output of the project is in demand and can find a ready market. (iv) The financing of the project is feasible, that the organisation and management needs.

   If the projects are made and the finally project is formulated in relation to absorptive capacity of the country.

   3. **Sound appraisal of the project :** It implies that the commercial, technical management financial and economic aspect of the project can satisfy various test of consistency and efficiency and therefore justify its worth to the society.

Points to be considered at the time of evaluation or Appraisal of project

(i) **Objective :** For what purpose project is formulated and implemented.

(ii) **Location :** Where do you want the project.

(iii) **With what resources :** Indicate the manpower, money, other input, management for expansions competition of project. All resources should be supplied as per requirements.

(iv) **When the project should be started :** There is need of time schedule for the start of the project.

(v) **Who should take this work :** By whom who should execute the project.

(vi) Why do you want a project.

If the opportunity cost high-project is preferred more alternative ways of the project.

Implementation of the Project

Certain precautions should be taken avoid poor performance of the project

Precautions : weakness in project

1. *Analytical weakness :* Means wrong or incomplete identification of the project and estimation of items of cost and benefits.

Analytical weakness due to :

(i) Unrealistic accounting prices

(ii) Persons involve and project analysis

(iii) Methodology adopted in analysis.

2. *Failure due to exogenous factors :* External factors, which affect the project, certain outward affects such as participation market imperfection, religions and customs.

Legal restrictions reacts on the implementation of the project :

3. *Incomplete identification of the project :* If identification is poor or incomplete, affect the project.

4. *Incomplete preparation of the project :* Formulation of project or project proposal under estimate, over estimate some factor are included or excluded.

5. *Climatic conditions :* When project is undertaken influence on working labour-affect completion of project. Efficiency of labour is reduced if adverse climate.

6. *Social and cultural environment :* Traditional and cultural activities of people participations understanding belief react with the project.

7.   Poor management on the part of executer or administrator of the project

Economic Evaluation of the Project

Accounting prices or shadow prices are considered in economic analysis. Economic analysis results in economic viability and technical feasibility.

Economic viability of the project is decided by different measures.

1.   Cash flow statements

2.   Pay back period (PBP)

3.   Net present worth or Net present value (NPV)

4.   Internal rate of return (IRR)

5.   Benefit cost ratio (BCR)

1.   **CASH FLOW STATEMENT** : also known as the sources and use of funds or flow of fund statements. Net cash flow required to estimate Net present worth or net present value (NPV). Viability is estimated by certain economic test - so to understand test one must be clear about some concepts such as cash flow, cash inflow, cash outflow.

$$\text{Net cash flow} = \frac{\text{Cash in flow}}{\text{(Income)}} - \frac{\text{Cash outflow}}{\text{(Expenditure)}}$$

Cash flows : It indicates stream of costs and benefits over a period of time. Basic principle of cash flow is that income and income and expenditure should be counted only at a time or at the same time. Therefore depreciation is not included in cash flow items. Total cash inflow includes (I) total cash available excluding borrowing (2) new operating loans (3) New MT/LT Loans.

Year	Cash outflow (cost)	Cash inflow (income)	Net cash flow
0	0	0	0
1	20	100	+80
2	40	200	+160
3	40	100	+60
4	40	100	+60
5	40	100	+60

To avoid double counting depreciation, interest are not included in cash flow because discounting takes care of this aspect. The net cash flow is estimated by subtracting cash outflow (cost) from the cash inflow (benefit) evaluation of project - Two methods.

1.   Discounting procedure-present value of future income determined.

2.   Compounding procedure is a process by which future value of present income can be determined.

Total cash outflow includes: (i) Total cash required (ii) Principal repayments (iii) Ending cash balance.

1. Net Present Worth or Value (NPW/NPV)

NPV of an investment is the discounted value of all cash inflow less all cash outflow of the project (Benefit - cost = Net cash flow) during its live period.

$$NPV = \frac{Rt - Ct}{(1-L)t} \quad \text{or} \quad \frac{NPV}{t=0} = \frac{Pt}{(1+r)^t}$$

Where,

Rt = returns (Benefits) in 't' time period

Ct= cost in 't' time period or cash outflow

L = discounted rate of interest

Rt – Ct = pt

Pt = Net present obtained

V = discounted rate of interest

Year	Net cash flow (Rs)	Discounting factor (10)	Present value
0	-1000	1	-1000
1	-100	0.87	-87
2	-300	0.756	227
3	-500	0.656	329
4	-500	0.572	286
5	-500	0.497	249
6	-500	0.432	216
		Additional NPV=	220

NPV > 0 – project undertake

NPV is positive – Accept project

NPV is Negative – Not undertake

NPV positive indicates that

1. Investor may be repay loan
2. It has some benefits equal to NPV

The important things in discounting procedure of NPV are

1. It allows for the compound interest payment at the rate of discounting rate.

2. The recovery of capital or principal payment over a period of time.

3. It indicates, it is wrong to include depreciation, interest payments, principal interest on a part of cash outflows or cost.

2. PAYBACK PERIOD (PBP)

Payback period can be defined as a time period within which initial investment of the project is recovered in the form of benefits; or

OR This is the length of time between the starting time of the project and the time when the initial investment is recouped in the form of yearly benefits.

Payback period – P = I/c

Where I = Initial investment C= yearly net cash in flow or Net cash income.

3. INTERNAL RATE OF RETURN OR YIELD (IRR)

IRR of a project is defined as the discount rate at which the present value of the investment is equal to zero (Symbolically)

$$NPV = \frac{Rt - Ct}{(lt\,I)t} = 0$$

It is that rate of return where the net benefit exactly PWP = PWC comes with the present value of the net cost

Where I = denots initial investment of project

Cj = Net incash flow = j = I, n = 5th period

N = life of the project

Rule : (i) Accept the project if the IRR or yield is higher than or equal to desired yield otherwise reject the project.

The project with higher IRR or yield will be preferred.

4. AVERAGE RETURN ON INVESTMENT (ARI)

Average return on investment (ARI) is computed in four steps.

Step-1. Find the total cash inflows over the project life.

Step-2. Subtract the initial investment from the total cash inflows and cost this total net income over the project life.

Step-3. Find the average annual income by dividing total net income by the project life in years.

Step-4. ARI is the ratio of average annual income to the initial investment.

5. BENEFIT COST RATIO

Benefit cost ratio of an investment is the ratio of discounted values off all cash inflows to the discounted values of all cash out flows during the life period of the project.

$$BCR = \frac{\text{Discounted all cash income inflows}}{\text{Discounted cost}}$$

$$BCR = \frac{\text{Present worth of benefits}}{\text{Present worth of costs}}$$

On the basis of categories:

(i) A project is worth considering if and only if payback period is not greater than the investor's desire. Maximum payback period.

PE < pd

(ii) NPV or NPW is positive than investment is profitable NPV > 0.

(iii) The investment of a project is economically viable if project guarantee and IRR which is greater than cost of borrowing or cost of capital → IRR > r

(iv) Similarly BCR > Unity then investment is profitable.

Economic Viability of Project

1. PE > pd E = Expected, d= desired

2. NPV > 0

3. IRR > r

4. BCR > unity

What are the methods for efficient evaluation/Appraisal of projects

There are four different types/methods of efficient evaluation (Appraisal) of projects.

1. Market appraisal/Evaluation

2. Technical evaluation

3. Financial evaluation

4. Economic evaluation

1. Market Evaluation

It is primarily concerned with two questions

(a) What would be the aggregate demand of the proposed product/ service in future

    (b)   What would be the market share of the project under evaluation.

          To answer the above questions the market analyst requires a wide variety of information and appropriate forecasting methods. The kinds of information required are

  (i)   Consumption trends in the past and present consumption level

  (ii) Past and present supply position

 (iii) Production possibilities and constraints

 (iv) Import and expert

  (v) Structure of competition

 (vi) Cost structure

 (vii) Elasticity of demand

(viii) Consumer behavior, intensions, motivations, attitudes, performances and requirements.

 (ix) Distribution of channels and marketing policies in use

  (x) Administrative techniques and legal constraints

2. Technical Evaluation

The evaluation of technical and engineering aspects of projects needs to be done continuously the technical analysis primarily consists with material. Production technique, location and site, plant capacity, machine and equipment, structure and civil works projects charts and layouts and works schedule.

3. Financial Evaluation

Financial evaluation seeks to ascertain whether the proposed project will be financially viable in the sense of being able to meet the burden of servicing debt and whether the project will satisfy the return expectations of those who provide capital.

The following aspects are considered while conducting financial evaluation are :

1. Investment outlay and cost of project
2. Means of financing
3. Cost of capital
4. Project profitability
5. Break even point
6. Cash flow of the project
7. Level of risk etc.

4. Economic Evaluation

Economic evaluation/analysis also refereed to as social cert benefit analysis is concerned with judging a project from the large social point of view on economic analysis the following questions are to be answered.

1. What are the direct economic benefits and costs of the project measured in the forms of shadow prices and not in terms of market prices.

2. What would be the impact of the project on the distribution of income in the society.

3. What would be the impact of the project on the level of saving and investment in the society.

TRADE

Meaning

Trade consists of the exchange of tools and services by two or more parties. Inter-regional trade is trade between different regions within the country. Whereas international trade is between different countries It be noted that the difference between inter-regional trade and international trade is only one of the degree.

Nature of Trade

Trade may be conducted on two bases :

Trade on Barter basis

In this case no money is exchange but goods and services owned by one party are exchanged for goods and services owned be another party. A more common means of trading is to exchange goods and service for money. Actuality money is commodity, which is accepted universally within a country in exchange for goods and services. Nations have different currencies, however, and the money of one nation is not normally accepted by another nation in exchange for goods arid service.

Geographically, the production of commodities is concentrated in the Specified areas and are specified in the production of the name commodity for, e.g., wheat in Punjab and Haryana, Rice in coastal area, Sugarcane in U.P. and western Maharashtra, Spices and condiments in Kerala, Jute in West Bengal, Tea and coffee in Assam and Oety in Chennai sate. These products or commodities had interregional as well as international trade. The commodities produced in a particular region are sold in the other region. The areas which arc deficit in production of one commodity exchange the commodities form the produced areas. This means that the concentration of production of certain commodities in agricultural area resulted wheat is known as a type of farming area. Areas tend to specialize in the production of particular commodities and trade with other areas.

Nations may also gain from Specialization of production and trade.

Inter Regional Trade

Inter-regional trade is the trade between different regions within the same country. This means trade from one region to another region. This trading activity is tackled by different states within the same country. Inter regional trade cannot take place from the country to another country. Trade among individuals within a nation is not usually conducted on barter basis. Rather, individuals sell goods and services to some people and purchase goods and services from others. The amount of goods and Services an individual can purchase over time depends upon the amount of goods and services he sells. The basic nature of the trade among nation is the same as trade between two individuals or two geographic areas within a country.

INTERNATIONAL TRADE

Trade in between two different countries is referred to as the international trade in other words international trade is concerned with transaction among nations.

The amount of goods a nation is able to purchase from other nations depends upon the amount of goods it can sell to other nations. The major difference is that two countries do not use the same money. The United Nations needs pounds, for example to purchase commodities produced in England. England on the other hand needs dollars to purchase commodities produced in the United States.

Trade Balance

Exports : Goods sold from one country to other countries are called as exports.

Imports : Goods purchased from other countries are called as imports.

When the amount purchased by a nation from other nations equals the amount sold to those nations the trade between nations is said to be in balance. Therefore, when exports equals imports we say that there is a trade balance between the nations.

Basically, trade between nations consist of the transfer of goods and services produced the granting capital loan the payment of interest on loans and dividends on investment in foreign countries and expenditure by tourists in foreign countries. Gold is internally accepted money. It is also some times exchanged directly for goods and services traded among nations. But this transfer normally do not take place since the monetary standard of the nation is based on the gold the nation is reluctant to purchase commodities involving transfer normally do not take place since the monetary standard of the nation is based on the gold the nation is reluctant to purchase commodities involving transfer of gold.

It is not necessary that the exports from any one country exactly balance the imports from another country for trade to continue to take place between two countries. What important is that the total amount of exports and import of a particular country over time must tend to balance.

It is the total exports that constitute receipts to a nation and determine its purchasing power, imports constitute expenditure. Although the exports and imports between two countries need not balance for trade between these countries to continue at a high level, it is necessary for the trade of a group of nations to balance. Suppose for example, that a country wishes to purchase goods produced in the United States and do not have the dollars to purchase those commodities. It may be able to obtain dollars by selling goods to another country which has sold goods to the United States for dollars. For example, suppose that England wanted to purchase commodities produced in United Stales but the England was short of dollars, she might obtain these dollars by selling British goods and services to African countries which have sold more commodities to the United States than they have imported from the United State.

FACTORS GOVERNING TRADE

1. **Mobility of factors of production :** Within the same country, labour and capital are more mobile than they are between different countries. Several courses are responsible for labour immobility like differences in languages, tradition, religion, custom, social and political life. Capital is more mobile than labour.

2. **Natural Endowments :** Difference in advantages of trade to different countries may arise because of natural courses like geographical and climatic conditions. These lead to territorial division of labour and localization of industries. For instance, some countries may have particular mineral resources like coal, iron ore, copper, etc., jute in Bengal. Either these advantages cannot be transferred to other countries at all or the cost of moving them is prohibitive.

3. **Human Capabilities :** Countries differ in human capabilities too. People in some countries are physically more study, whereas, in others they are intellectually superior. Some have greater skill and dexterity and other excel in spirit of enterprises and organizational ability.

4. **Stock of Capital :** Some countries posses large stock of capital goods like the UK and USA and other like India suffers from deficiency of

capital. This makes a great difference in the types of goods produced in different countries.

5. **Political Sovereignty :** In international trade, certain problems arise out of fact that countries are independent sovereign states and can pursue independent policies with respect to the movement of goods, wages and prices, fiscal matters, banking law, foreign loan etc. Several kinds of restrictions are placed on the movement of goods beyond their frontiers by the states.

6. **Currency System :** Different countries have different currency system. This hampers smooth flow of trade as between one country and another. The number of foreign exchange problems arise in foreign trade which is non-existent in internal trade.

7. **Separate Markets :** There are cultural distinctions between markets. The national markets are frequently separated from one another e.g., the British use Right Hard drive car, whereas the French use the Left Hand drive cars. Thus markets for automobile are effectively separate. Markets are also separated by language, customs, usage habits, tasters, standards differ some goods are designed in inches, feet and short tons and other in metric measurements.

8. **Economic Nationalism :** Different countries have their separate income life. "Along with political independence has grown demand for economic reliance, self esteem expressed largely in plant and hopes for economic development". The national, units have been striving for increasing consumption, production, capital formation etc. thus political and economic nationalism rising specifically in newly independent countries widening the differences between international and inter regional trade.

9. **Trade and Exchange Controls :** Trade and exchange control are instituted by almost all modern states, which construct the movement of goods and services from one country to another.

INDIA'S TRADE POLICY : A PRESENT SCENARIO

In advanced countries trade policy is used for :

(1) To restrict their imports and provide a sheltered market for their own industries for rapid development.

(2) Promote their exports so that their expanding industries could secure foreign markets.

Hence in advanced countries trade policy played very important role for their development. In India, before independence we could not have clear trade policy but there was some restriction on imports known as discrimination protection was adopted since 1923 to protect a few domestic industries against foreign competition. After independence trade policy become a part of economic policy of development of India.

MAIN FEATURES OF IN TRADE POLICY

(A) **On the import side :** India has been in a disadvantageous position as compared to advanced countries, which are capable of producing and selling almost every commodity at low prices. This means that that to industry in India protected against foreign competition Thus it was essential to protect domestic industry and to promote industrial development. Since, independence, Govt. of India has broadly restricted foreign competition through a judicious use of import licensing, import quotas, import duties and in extreme canning import of specific goods. The Mahalbois strategy of economic development in second plan called for.

   (a) Banning or keeping to the minimum import of non-essential consumer goods.

   (b) Comprehensive control of various products import, liberal import of machinery, equipment and other developmental goods to support heavy industries.

   (c) Favourable climate for the policy of import substitution.

(B) **On the export side :** Expansion of export is essential for minimizing the dependence of country on foreign countries in India sonic products/goods no adequate market hence for surplus production trade there is necessity to search up trading institutions, and through fiscal and other incentives. Vigorous export promotion was taken place in II nd plan to earn foreign again emphasized because of mounting debt services obligation and goal of self-reliance.

PHASES OF INDIA'S TRADE POLICY

Five distinct ph1ases in India's trade policy can be noted :

   (1) The first phase pertains to the period 1947-48 to 1951-52.

   (2) Second phase covering the period 1952-53 to 1956-57.

   (3) Third phase after 1956-57 to June 1965.

   (4) Fourth phase started after devaluation o the Rupee in June 1966.

   (5) 1975-76 to 1990.

   (6) 1991 onwards.

(A) **During first phase (1947-48 to 1951-52)** : India could have liberalized imports but on account of the restriction placed by the U.K. our balance payments with the dollar area was heavily adverse. Efforts were made to screen imports from hard currency area and boost up exports to this area so as to bridge up gap. Devaluation of money was necessary in 1949. Imports and exports have restrictions during this phase.

(B) **During Third phase (1956-57 to June 1965)** : The trade policy was re-oriented to meet the requirements of planned economic development. A very destructive import policy was adopted and import controls further screened the list of import goods. On the other hand, vigorous export promotion drive was launched. Export of traditional items as well as new items was promoted similarly, import substitution industries encouraged for reducing dependence on foreign countries. It was in this period that India's trade policy was thoroughly reviewed by the Mudaliar committee (1962).

(C) **During fourth phase 1966 onwards** : The fourth phase started after the devaluation of the rupee in June 1966. During this period trade policy attempted to expand exports and strongly liberalized imports too. Accepted recommendations of the Mudaliar committee included increased allocation of raw materials to export-oriented industries, income tax relied on export earning, export promotion through import entitlement, removal of disincentives, setting up of Export Promotion Advisory Council, a ministry of international trade etc., when then these took devaluation of the rupee for check imports and boost export this period continued till 1975-76.

(D) **During Last phase (1975-76 onwards)** : The Govt. adopted policy of import liberalization with a view to encourage export promotion to augment domestic supply of essential goods and to check rise in price level. The Import- Export policy of Indian Govt. attempted to achieve following objectives.

   (I)   To provide further impetus to exports.

   (II) Give support to growth of indigenous industry.

   (III) To facilitate technology up-gradation with promotion of export and energy conservation.

   (IV) To provide stimulus to savings in imports.

Thus it is clear that purpose of trade policy has been to stimulate economic growth and export promotion via import liberalization.

Import liberalization along with export promotion of a time when (i) prices of imported goods were rising much faster. (ii) Foreign markets for Indian goods were depressed resulting in huge adverse balance of trade and payments from

1970-80 onwards. Tendon committee (1981) and IMF loan (1981) also recommended that export promotion and no restrictions on import. While framing the export-import policy (1985) the govt. was guided by Abid Husen.

The commerce minister Mr. P. Chidambaram announced a major over of trade policy on July 4, 1991.

    (i)   Suspension o cash compensatory support.

    (ii)   An enlarged and uniform Rep. Rate of 30 per cent of food value.

    (iii)   Abolition of all supplementary licenses except in the case of small scale sector and producers of life saving drugs/equipments.

    (iv)   Abolition of unlisted OGL and Removal of all import licensing for capital goods and raw material except for a small negative 1 years.

EXPORT PERFORMANCE OF AGRICULTURE IN INDIA

Agriculture, as an important sector in Indian economy, facing various challenges in globalization. It is required to prepare ourselves in appropriate manner, knowledge, investment incentive, information and internationally of the farm attributes essentially being developed this purpose. Indian agriculture has transformed from food shortage to self-reliance and surplus owing to technological break through as well as policy and programmatic initiatives of the government and response of our farming community to become active agent of development design.

India's export during 1970 was Rs. 8770 crores which increased to Rs. 293820 crores during 2005-06 at current prices. This showed that India's export multiplied by 41.26 times over 1970. In term of share in total world export, the export share of India in current year is 0.87 per cent and it was 0.67 per cent during the year 2005. In short, India's share is hardly less than one per cent of the world trade.

There is a large scope to increase the share of export of agricultural product through getting benefit of WTO and new technologies in agriculture. There is large scope for the export of value added products in horticulture. In this view, the study on the export performance of agriculture in India is conducted.

Share of agricultural export and import in India's total export

India has been importing foodgrains for quiet some time after independence. It has also been exporting agriculture products their by helping the country not only to pay for food import but for other imports which includes capital food also. It is noted that, the total export of the country during the year 1990-91 was to the extent of Rs. 325272 millions which has been significantly increased to Rs. 4548000 millions in 2006. Though, over the year agriculture export has increased many fold but the share of agriculture export has decreased from 18.49 per cent in 1991 to 10.95 per cent in 2006. Similarly, the total import were to the tune of Rs. 12059 million in 1991 which has increased to Rs. 210255 million in 2006. The agriculture export and import of country showed an increasing trend in absolute terms but had declining trend in the percentage share in the total export and import of country. The percentage share of agricultural import had increased from 2.79 per cent to 8.17 per cent during the period from 1990-91 to 1998-99 then declined upto 3.33 per cent in 2005-06.

Share of agriculture import/export to National import/export (Value : million Rs.)

Year	Import			Export		
	Agriculture	Total (National)	% Agril to total (National)	Agriculture	Total (National)	% Agril. total (National)
	Value	*Value*		*Value*	*Value*	
1990-91	12058.60	431708.20	2.79	60127.60	325272.80	18.49
1995-96	58901.00	1226781.40	4.80	203977.40	1063533.50	19.18
2000-01	120862.30	2283066.40	5.29	286573.70	2013564.50	14.23
2005-06	210255.40	6305267.70	3.33	498029.20	4547999.70	10.95
CGAR	141.61	161.00		182.00	165.00	
1990-91 to 2005-06						

Source : Data Yearbook 2007.

EXPORT OF HORTICULTURE PRODUCTS

It is observed from table that, India's export of horticultural products increased by 28 per cent during last three years. The export earning of horticultural product valued to the tune of Rs. 12069 million in 2002-03 increased to Rs. 14627 million in 2006, i.e., it was increased by 48 per cent during last three year. The export dried and processed vegetable increased by (25%), fresh onion (20.5%), mango pulp (10%) followed by other processed fruits and vegetables (9%), floriculture and other fresh vegetables (7%), other fresh fruits (5%), fresh grape and pickles and chutney (4%), etc in last three year.

Export of horticultural products (export share and value)

(Quantity : t, value : in million Rs.)

Products	Quantity			% share of qty. of 2004 -05	Value			% share of value of 2004-05
	2002-03	*2003-04*	*2004-05*		*2002-03*	*2003-04*	*2004-05*	
Floriculture	N.A.	N.A.	26262	-	1653.9	2495.5	2109.9	6.9
Fruits and vegetable seeds	10658	5170	6307	68.97	1009.7	536.0	629.4	2.1
Fresh onions	588712	859939	833209	29.34	3618.0	7158.7	6210.9	20.5

Contd...

Other fresh vegetables	183019	188321	181957	-0.58	2876.4	2522.8	2243.9	7.4
Walnuts	7631	6418	5674	-34.49	1212.3	1014.3	928.3	3.1
Fresh grapes	25681	26784	35936	28.54	1101.5	1058.9	1106.7	3.6
Fresh mangoes	38003	60551	52382	27.45	841.9	1105.2	869.5	2.9
Other fresh fruits	90609	149294	131542	33.68	1217.4	1712.7	1640.0	5.4
Dried and preserved vegetables	216640	211160	351034	38.29	5610.3	5204.9	7657.5	25.2
Mango pulp	96107	89515	90989	-5.63	2970.1	2419.9	3008.6	9.9
Pickles and chutney	56384	63053	67193	16.09	1541.6	1197.5	1205.8	4.0
Other process fruits & vegetables	54793	66070	80761	32.22	1947.3	2435.8	2755.3	9.1
Total	423925	429798	589977	28.15	12069.3	11258.1	14627.2	48.2

Source : Data Yearbook 2007.

India's export and import of selected agricultural commodities

The commodities exported from India broadly fall in three categories :

(1) Traditional export items, these products are cashew nuts/shelled, castor oil, coffee, raw cotton, cotton waste, fruits, spices, sugar and molasses, tea and tobacco unprocessed, (2) Non-traditional items but uncertain, these items are raw jute, raw wool, gum, resin and lac, essential vegetable oil and non-essential vegetable oil (including castor oil), (3) Non-traditional items with good prospects these are floricultural products, HPS groundnut, oil meal, meat and meat preparation, processed fruit and juice, processed vegetable, sesammum and niger seed, shellac, wheat and rice. India's exported and imported agricultural value to tune of Rs. 664566 million and Rs. 458268 respectively in 2006.

The commodity, wise export increased by about 28.36 per cent and import by 40.23 per cent during year 2002-03 to 2005-06.

EXPORT OF WHEAT, RICE, GROUNDNUT AND SPICES

The export of wheat, rice, groundnut and spices during 1990-91 to 2004-05. It showed that the export of all five crops i.e. wheat, rice (Basmati), rice (other), groundnut and spices have recorded positive and significant CGAR. The export of wheat has increased from Rs. 29.23 crores in 1990-91 to Rs. 2349.37 in 2003-04, then slightly declined to Rs. 1448 crores in 2004-05. The CAGR is 64%, which is positive and significant. The export of rice (Basmati) and rice (other) increased during this period. The export of rice (Basmati) was Rs. 439.95 crores in 1990-91 and went upto Rs. 2742 crores in year 2004-05. Similarly the rice other than Basmati increased from Rs. 256.41 crores in 1991-92 to Rs. 3899 crores in 2004-05. CAGR of rice (Basmati) is 185 % and rice (other) is 22% which is positive and significant. The export of groundnut was Rs. 58 crores in 1990-91 and it increased upto Rs. 502 in 2004-05. The CAGR is 6.0%, which is positive and significant. The export of spices was Rs. 242.11 crores in 1990-91 and increased upto Rs. 2200 crores in 2004-05. The CAGR has been 118% which is positive and significant.

Export of wheat, rice, groundnut and spices (Rs. crores)

Year	Wheat	Rice Basmati than Basmati	Rice other	Groundnut	Spices
1990-91	29.23	439.95	0	58	242.14
1991-92	126.98	499.18	256.41	7.34	380.96
1995-96	366.76	850.67	3717.41	230.69	804.43
2000-01	415.07	2145.94	777.26	316.4	1833.53
2001-02	1330.21	1842.77	1331.37	250.94	1940.55
2002-03	1759.87	2058.47	3772.77	178.3	2086.71
2003-04	2349.37	1990.92	2175	179	1911.61
2004-05	1448	2742	3899	502	2200
CGAR	64.00	185.00	22.00	6.00	118.00

Source : Various Government publication.

Import of wheat, rice, oilseed, veg. oil and pulses (Rs. crores)

Year	Wheat	Rice	Oilseeds	Veg. oil	Pulses
1990-91	24.19	39.19	6.42	322.22	473
1991-92	0	10.94	9.65	247.79	255
1992-93	710.06	73.32	10.64	166.88	315
1993-94	125.62	55.26	6.98	166.63	567
1994-95	0.83	8.55	5.35	624.24	573
1995-96	10.39	0.05	36.17	2261.93	685
1996-97	403.76	0.02	4.7	2929.19	890
1997-98	988.98	0.06	2.47	2764.67	1194
1998-99	1164.78	5.4	8.52	7588.93	708
1999-00	774.35	29.95	15.42	8046.05	354
2000-01	2.87	17.79	7.21	5976.53	498
2001-02	0.84	0.07	1.34	6464.97	3160
2002-03	0	1.09	11.49	8779.64	2565
2003-04	0.2	0.27	13.89	11683.24	84.87
2004-05	0	0	28.41	11076	1777.5
2005-06	0	0	47.24	8716.8	2346.9
CGAR	-65.00	-26.00	24.00	83.00	57.00

Source : Various Government publication.

India's import : It is related to import depicts that import of wheat is showing remarkable variation from Rs. 24.19 crores in 1990-91 to highest Rs. 1164.78 crores in 1998-99 but then again it declined upto Rs. 0.84 crores in 2001-02. Therefore CAGR during the period was negative at 65% which is insignificant. Import of rice also declined from Rs. 39.19 crores in 1990-91 to Rs. 5.4 crores in

1998-99 and in year 2003-04 it is 0.27 crores. The CAGR is 26% and it is insignificant while, import of oil seeds and vegetable oils have continuously increased from Rs. 6.42 crores in 1990-91 and reached 47.24 crores in 2005-06 and 322.22 crores in 1990-91 to 11683.24 crores in 2003-04 and then declined to 8716.80 crores in 2005-06. The CAGR for both oilseeds and vegetable oil is 24 and 83% respectively which is positive and significant. The import of pulse have continuously increased from Rs. 473 crores in 1990-91 and to 3160 crores in 2001-02 and then declined to 2346.9 crores in 2005-06.57 The CGAR is 57%. and is positively significant.

The share of agricultural export to India's total export is continuously declining over the year from 18.49 to 10.95 per cent during 1991 to 2005-06. The percentage share of Agricultural import to total import had increase from 2.79% to 8.17% during the period from 1990-91 to 1998-99 then decline upto 3.33% in 2005-06. The export of wheat, rice, groundnut and spices during 1990-91 to 2004-05. It showed that the export of all five crops, i.e., wheat, rice (Basmati), Rice (other), groundnut and spices have recorded positive CAGR and is significant. The import of wheat and Rice during 1990-91 to 2004-05. It showed that import of two crops have negative CGAR at it is non-significant. While import of oil seed, vegetable oil and pulse have positive and significant CGAR. Though India is second largest producer of fruits and vegetables in the world. It's share in the world export is just 1.06 per cent in 2003-04 with huge production. There is potential to increase the export of horticultural products. There is need to upgrade the grading and marketing acts alongwith other related acts like land ceiling, local taxes and duties and various other agricultural regulations. The improved functioning of APMC, integration of various marketing activities, the co-operative spirit, private trader's skill and government watchful supervision will definitely improve the status of agricultural marketing in domestic and international market. There is large scope to increase the share of export of agricultural product through getting benefit of WTO and new technologies in agriculture. There is large scope to export of value added products in horticulture.

Advantages and Disadvantages of International Trade

Benefits from foreign trade form specialization on the basis of comparative development. It prevents monopolies. It also facilitates international payments. The gain from international trade can be broadly classified into static gains and dynamic gains static, gains arise form optimum use of country factor endowment or resources in men money and gains material so that national input is maximized resulting in increase in social welfare. Dynamic gains, on the other hand refer to those benefits which promote economic growth of participating countries.

Static Gains

Static gains results form the operation of the theory of comparative cost in the field of foreign trade. Acting on this principle the participating countries are able to make optimum use of their resources or factor endowments so that the national output is greater that is otherwise would be. This raises the level of social welfare in the country. Utility or welfare can be measured by indifference curve.

Dynamic Gains

Specialization by different countries in producing commodities for which they are best fitted, according to the theory of comparative cost, result in the large volume of the production and improves the productivity. This obviously, promotes economic development. The extension of international trace accelerated economic growth in participating countries. Foreign trade promoted economic development in the following different ways.

1. The under developed countries are enabled by foreign trade to obtain in exchange for goods capital equipments, machinery and raw material which are highly useful in accelerating the rate of economic growth.

2. Besides raw material, machinery and capital equipments, international trade enables, a country to import technical knowhow technical kilts, managerial talents and entrepreneurship through foreign collaborations.

3. The existence of a large volume of foreign trade serves as a guar meet for the payment of interest and the principle.

Consequences of International Trade

Advantages and disadvantages of international trade are summarized as :

1. **Equalization of commodity prices :** One direct effect of international exchange of goods to equalize the prices of similar goods in the trading countries. Absolute equality is however not possible. There must remain some difference in the prices of the internationally traded goods to compensate for transport cost and other incidental expenses.

2. **Equalization of factor prices :** Commodity prices ultimate analysis depends on the factor prices. Hence equalization of commodity prices must tend to equalize factors prices. The prices of relatively scare factors will fail since the good in which they are used will be imported and demand for such factor will diminish. On the other hand, there will be greater demand for relatively cheap and abundant factors for the goods in which they enter will be exported. Their price will, therefore, tend to rise, But complete equalization cannot he expected.

3. **Equalization, Distribution of Scarce Material :** Nature has blessed some countries with some rare material like oil in Arab countries. These scare materials are equitably distributed among the countries of the world through international trade.

4. **Effect on factor supplies :** It is also possible that the fall in price of scarce factor may lead to contraction of their supply and rise in the prices of the abundant factor may result in the extension their supply. This may accentuate differences in factor supply of the two countries.

5. **Specialization and international Division of Labour :** It will be of mutual advantage to the trading country. Not only their standard of living will rise but they will also progress economically and industrially. The productive resources will be put to optimum use. Disadvantages arising from uneven distribution of factor endowments will disappear although. It must be admitted that the gain occurring from international trade will not be distribute equally. Advanced countries will gain relatively more than the under developed countries.

6. **International Trade affects People's taste and desires :** This means that demand for certain goods will increase and demand for some new goods will arise. In this manner, international trade will affect both the volume and the nature of demand.

Barriers of Foreign Trade

1. **Import prohibition :** Sometimes import of certain commodities is prohibited by low or allowed only under defined conditions e.g., "Sanitary regulation". The USA once excluded beef from a certain region of Argentina where foot and mouth disease had attacked cattle. Later on embargo was extended to the whole Argentina.

2. **Exchange Control :** Exchange control implies govt. interference with the buying and selling of foreign exchange. In this way, foreign trade is curtailed and driven into channels. Govt. may "allit" exchange or ration it so that importers can buy only a limited amount of goods in foreign countries or they may "block" exchange, e.g., America exporting goods to Germany may be required to use the marks exchange thus obtained in purchasing goods in Germany.

3. **Custom Duties :** Custom duties consist of import and export duties on goods entering into or going out of the country respectively, custom duties or tariffs may have either revenue or protecting aim, e.g., to project cotton industry an export duties on raw cotton may be imposed to cheapen it for the home manufacturer or an import duties on the cotton manufacture may be levied. This duty has greater continuity. While the revenue duly is for revenue primarily and is levied for the financial year. It may he revised discontinued in the next year.

4. **Preferential Treatments :** Sometimes discrimination is the rate of duties with regard to different countries, e.g., India gave preferential treatment to certain British goods under the Agreement of 1932, such arrangement

curtail international trade and leads to the development of trade blocks. Moreover, countries goods pay higher duties may retaliate and impose high duties on the discriminating country in return.

5. **Quota Restriction :** There are two kinds of quotas 'Custom Quota' and 'Import Quota'. The first type allows a certain amount of commodity at a favorable duty; beyond this the normal duties are charged. The import quota has more serious effect on the trade. Here an arbitrary limit is set, beyond which imports a given period are not allowed. Due to quota system home market is isolated from the world market. Prices may be falling outside but they will have no effect on the home market because beyond a certain limit they cannot import.

6. **Import Licenses :** Under this system, the govt. does not allow import of certain goods without a license being obtained by importer. In this very imports he cut down and certain goods discriminate against.

7. **Import Monopolies :** The govt. may make the import of goods as state monopoly as Russia does and thus reduce import or discriminate against certain countries.

REFERENCES

1. Acharya, S.S. and Agarwal, N.L. (2004). *Agricultural Marketing in India*. Oxford and IBH Publishing Co., Pvt. Ltd. New Delhi.

2. Subba Reddy, S. and Ragha Ram, P. (1996). *Agricultural Finance and Management*. Oxford and IBH Publishing Co., Pvt. Ltd. New Delhi.

3. Ghosh, G.N. and Raj Ganguly (2008). *Development Challenges of Indian Agriculture*. FAO Indian Working Documents.

4. Berkely, Hill. (2006). *An Introduction to Economics*. 3rd End. Athenaeum Press. UK.

5. Lekhi, R.K. and Singh, J. (2006). *Agricultural Economics*. Kalyani Publishers, New Delhi.

6. Bhatia, G. (2007). *Agribusiness Management*. Mittal Publications, New Delhi.

7. Broadway, A.C. and Broadway, A.A. (2002). *A textbook of Agri-business Management*. Kalyani Publishers, New Delhi.

8. Mamoria, C.B. and Joshi, R.L. (1975). *Principles and Practices of Marketing in India*. Kitab Mahal, New Delhi.

9. Kotlar, P. and Armstrong, G. (1997). *Principles of Marketing* VIIth edn. Prentice-Hall of India Pvt. Ltd., New Delhi.

10. Acharya, S.S. and Agarwal, N.L. (1999). *Agricultural Marketing in India* 3rd edn. Oxford and IBH Publishing Co., Pvt. Ltd. New Delhi.

KEY NOTES ON STATISTICS

1

TERMINOLOGY

Terms	Terminology
1% level of significance	It shows that 99% confidence that the decision made is correct.
5% level of significance	It shows that 95% confidence that the decision made is correct.
Arithmetic mean	Arithmetic mean is of series of items to obtain by adding the values of dividing it by number of items.
Central tendency	Tendency of items to concentrate at central value is known as central tendency.
Coefficient of determination	The coefficient of determination means square of correlation. It is a ratio between explained variance to the total variance.
Coefficient of variation	Coefficient of variation is a ratio between standard deviation with arithmetic mean in percentage.
Correlation	The statistical tool with the help of which these relationships between two or more than two variables are studied is called correlation.
Data	Data is a group of observations, classifications of data grouping of data according to their resemblance affinities and common properties of factors.
Degree of freedom	The degree of freedom is obtained by deducting one from the total number of items of the number of samples.
Geometric mean	Geometric mean is defined as the nth root of the product of 'N' items of series. If there are two items, we take square root.
Harmonic mean	It is defined as the reciprocal of the arithmetic mean of the reciprocal of the individual observations.

Terms	Terminology
Karl person's correlation coefficient	Ratio of sum of the product of deviation of two variables from their arithmetic means to the product of number of pair of observations and corresponding standard deviation of variables is defined as Karl person's correlation coefficient.
Large sample	A sample in which number of observations is more than thirty [30].
Level of significance	The accuracy with which the decision of rejecting the null hypothesis made is known as level of significance.
Line of regression of X on Y	X = a+by
Line of regression of Y on X	Y = a+bx
Mean deviation about arithmetic mean	It is defined as arithmetic mean of the deviation of the individual item in the data from central value.
Mean deviation about median	When deviation from individual item taken from median the mean deviation known as the mean deviation from median.
Median	When data is arranged in ascending or descending order the value of central item is known as median of that series or data.
Mode	Mode is defined as a value of item, which is, repeated highest number of items in the data.
Multiple correlation	Correlation coefficient between a dependant variable and a group of independent variables is called multiple correlation coefficients.
Multiple correlation coefficient	It is common to add subscripts designating the variables involved.
Multiple regression	When comparison between two or more linear regression is known as the multiple regressions.
Null hypothesis	Any hypothesis regarding the population parameter in the form of statement of equality, which can be either nullified or disproved, is known as null hypothesis.

Terms	*Terminology*
Partial correlation	The relation between a dependent variable and one particular independent variable when all other variables involved are kept constant or effects are eliminated.
Partial correlation coefficient	It provides a measure of the relationship between the dependent variable and other variables, with the effect of the rest of the variables eliminated.
Partial regression	When comparison between two linear regressions at a time is known as partial regressions.
Population	A group of observations made out individual living or non-living possessing some common character, which varies from individual to individual, is known as population.
Population parameter	Any summering quantity calculated from population is known as population parameter.
Probability of an event	Probability of an event is a ratio between numbers of success of an event to the total number of trials.
Range	Range is defined as difference between the value of highest item and the value of lowest item included in the distribution.
Regression coefficient	It is defined as the rate of change of dependant variable per unit change in the independent variable.
Sample	A part of population is properly selected to study the population and to present the population is known as sample.
Sampling distribution	The distribution so formed of all possible values of a statistic is called the sampling distribution.
Small sample	A sample in which number of observations is less than thirty [30].
Standard deviation	It is defined as the root mean square deviation from an arithmetic mean and it is denoted by "s".
Standard error	The standard deviation of the sampling distribution is called the standard error.

Terms	Terminology
Statistical hypothesis	It is mathematical statement, which can be tested with the help of statistical distribution.
Statistical inference	Any conclusion drawn about population parameter with the help of sample statistical is known as statistical inference.
Statistics	Any summering quantity calculated from sample for estimating the population character is known as statistics.
Statistics	The science, which deals with the collection, presentation, analysis and interpretation of numerical data.
Variance	Variance is defined as the square of standard deviation [s^2].

2

SYMBOLS / ABBREVIATIONS

Symbol	Symbol Name
A	Assumed Mean
b_I or b_{xy}	Regression Coefficient of X on Y
b_2 or b_{xy}	Regression Coefficient of Y on X
C	Common factor
c. f.	Cumulative frequency
Coeff.	Coefficient
d	(X-A), i.e., deviation of a value X from an assumed mean
d'	(X -A)/C i.e., deviation of a value X from an assumed mean taking a common factor
I d I	deviation of items from median (or mean) ignoring sign
d.f.	degrees of freedom
$D_1, D_2, D_3,$ etc.	1st, 2nd, 3rd, deciles etc.
E	Expected value
f	Frequency
f_0	Observed frequency
f_e	Expected frequency
$f_1, f_2, f_3,$ etc.	Frequency of first, second, third classes. etc.
G.M.	Geometric Mean
H.M.	Harmonic Mean
i	Class Interval of a class
Log	Logarithm
l	Lower limit of a class
U	Upper limit of a class
m	midpoint of a class

Symbol	Symbol Name
Med	Median
Mo	Mode
M.D.	Mean Deviation
N	Number of observations or sum of frequency, i.e., Σf in case of a frequency distribution
O	Observed value
p	Probability of happening of an event
P_1, P_2, P_3, etc.	First, second, third, percentiles etc.
P_o	Price in the base year
P_1	PI=price in the current year
Q_1, Q_2, Q_3, etc.	First, second, third, quartiles etc.
q_0	Quantity in the base year
q_1	Quantity in the current year
Q	Yule's Coefficient of Association
Q.D.	Quartile Deviation
r	Coefficient of correlation
r_k	Rank Correlation
r_c	Correlation by concurrent deviation method
r^2	Coefficient of determination
SK	Coefficient of Skewness
C.V.	Coefficient of Variation
W	Weights
X_1, X_2, X_3, etc.	Individual observations of the variable X
X	Arithmetic Mean
x	(X-X) i.e., deviations of items from actual mean
X'	(X-X)/C, i.e., deviations of items from mean taking a common factor, i.e., step deviations
X_{12}	Combined mean of two series or groups
$X_{11\backslash 23}$	Combined mean of three series or groups
X_w	Weighted arithmetic mean
Y_1, Y_2, Y_3, etc.	individual observations of the variable Y
2	Arithmetic mean of Y series
y	(y-y)
Z	X-X/σ (standard normal variate)
Ó	Summation 'or the sum of'

3

DISTINGUISH/COMPARISON

Compare between Qualitative and Quantitative clarification

Qualitative classification	Quantitative classification
1. In qualitative classification data are classified on the basis of some attribute or quality.	1. Quantitative classification refers to the classification of data according to some characteristic that can be measured
2. Ex. Sex, colour of hair, literacy, religion, blindness etc.	2. Ex. Height, weight, length etc.
3. No further classification of qualitative data.	3. Quantitative data can be further classified as follows, discrete data and continuous data.

Differentiate between Discrete and Continuous Variables

Discrete variables	Continuous variables
1. Discrete data counted in full number	1. The continuous data varies fractional values
2. Ex. Numbers of student in one class, Number of seeds/pod, Number of earheads/plant	2. Ex. Height of plant, weight of seeds, length of leaves

Differentiate between Arithmetic mean and Standard deviation

Arithmetic mean	Standard deviation
1. Arithmetic mean is of series of items to obtained by adding the values & dividing it by number of items	1. Standard deviation is the root mean square deviation from an arithmetic mean
2. Best measure of central tendency	2. Best measure of the dispersion
3. Arithmetic mean is denoted by X.	3. The standard deviation denoted by σ
4. It is use for further algebraic calculations.	4. It is prominently used for all statistical calculations
5. It is very easy to calculate	5. It is difficult to calculate
6. It is based on all items	6. It gives more weight to extreme item & less weight to small items
7. It does not vary too much	7. It is too much vary from sample to sample.

8. Formula	8. Formula
$X = x_1 + x_2 + x_3 + \text{------} + x_n/n$ n = number of items	$\sigma = \Sigma\,(xi - x)^2/N$ N = number of items

Differentiate between Standard deviation and Coefficient of variation

Standard deviation	*Coefficient of variation*
1. S.D. is defined as the root mean square deviation from an arithmetic mean & it is denoted by σ	1. C.V. is a ratio of S.D. and mean in percentage
2. The standard deviation is absolute measure of scatter of all above arithmetic mean.	2. The coefficient of variation is not absolute measure of scatter of all items above arithmetic mean.
3. The standard deviation is not useful for comparing the variability of two or more characters which are different in the nature.	3. The coefficient of variation which is used for comparisons of characters different in nature.
4. Standard deviation having same unit in which the items (xi) is measured.	4. Coefficient of variation is expressed in percentage. It is unit less quantity.
5. Formula $\sigma = \Sigma(xi - x)^2/N$	5. Formula C.V. $= \sigma\,/x \times 100$

Differentiate between Arithmetic mean and Median

Arithmetic mean	*Median*
1. Arithmetic mean is of series of items to obtain by adding the values and dividing it by number of items.	1. When data is arranged in ascending or descending order the value of central item is known as median of that series or data.
2. Influenced by extreme items in the data	2. Not influenced by value of extreme items in the data
3. It is calculated value & it does not depend upon position	3. It is depend upon the position
4. It does not gives central value or central item	4. It gives the value of central item when data arranged in ascending or descending order.
5. Arithmetic mean do not divide data in to equal parts	5. The median divide the data into two equal parts.
6. The values of arithmetic mean does not determined by graphically	6. Value of median determined by graphically
7. While calculating arithmetic mean all items in the data are taken into consideration	7. While calculating median all items in the data are not taken into consideration.

Contd...

8. Such type of arrangement do not required to the calculating arithmetic mean.	8. For calculating median data should be arranged in order
9. Arithmetic mean useful for further algebraic calculation	9. Median is not useful for further algebraic calculation.
10. Arithmetic mean is best measure of central tendency.	10. Median is not best measure of central tendency

Differentiate between Correlation coefficient and Regression coefficient

Correlation coefficient	*Regression coefficient*
1. Correlation coefficient gives degree & direction of relation ship between two variables.	1. Regression coefficient gives nature of relationship between two variables so that the dependent variable can be estimated.
2. In correlation study cause and defects may not be define.	2. In regression analysis cause & effects are clearly defined
3. For example, Correlation between grain yield and rainfall.	3. For example, the grain yield dependent upon rainfall
4. rxy=rxy correlation coefficient is symmetric in x and y	4. byx ≠ bxy Regression coefficient is not symmetric in x and y.
5. rxy = ruv It is independent on change in origin	5. bxy = buv It is independent on change in origin
6. rxy is dependent of change of scale.	6. bxy is dependent on change in scale.

Differentiate between Scatter diagram and Correlation coefficient methods

Scatter diagram method	*Correlation coefficient method*
1. The scatter diagram method gives the correlation by diagrammatically & not by the mathematically	1. The correlation coefficient do not give diagrammatically correlation but gives by the mathematically
2. It gives direction of correlation i.e. +ve or −ve.	2. It gives degree & direction both.
3. Does not give degree of correlation i.e. rxy cannot be calculated.	3. rxy can be calculated by the correlation coefficient method.
4. It is dependent upon the direction	4. It is independent on change in origin and scale.
5. It gives following different values at different directions. When +ve correlation $0 < rxy < 1$ When perfect correlation $rxy = 1$	5. Correlation coefficient is symmetric in x and y and it is independent on change of scale and origin.

Contd...

When −ve correlation
$1 < rxy < 0$

When perfect −ve
correlation $rxy = -1$

When correlation not +ve
nor −ve then $rxy = 0$

Differentiate between Karl Pearson and Spearman's rank coefficient methods

Karl Pearson coefficient method	Spearman's rank coefficient method
1. Ratio of sum of the product of deviation of two variables from their arithmetic means to the product of number of pair of observations & corresponding standard deviation of variables is defined as Karl Pearson's correlation coefficient.	1. It is the ratio between six into addition of all ranks diff. square to the cube of number of observations minus number of observations. These all subtracts from the one. The remaining value is known as the correlation coefficient of separation's rank.
2. It is denoted by rxy	2. It is denoted by $\int_o xy$ or rs
3. Useful for quantitative data	3. Useful for qualitative data
4. It gives degree and direction of correlation	4. It gives degree and direction of correlation.
5. Such type of restriction is not required for this correlation	5. It is useful when data is not normally distributed
6. This method uses the actual data & not their rank.	6. This method does not use the actual data but their ranks are used.
7. This method can be used for grouped frequency distribution.	7. This method cannot use for grouped frequency distribution.
8. Not such type of problem occurs in this method.	8. When N greater than 30 calculations become tedious i.e. very difficult.
9. Karl Pearson method is difficult to understand & complicated to calculation.	9. Rank correlation coefficient method is very simple to understand & easily calculated.

Differentiate between 5% and 1% level of significance

5% level of significance	1% level of significance
1. It shows that 95% confidence that the decision made is correct.	1. It shows that 99% confidence that the decision made is correct.
2. It shows that 5% confidence that decision made is not correct.	2. It shows that only 1% confidence that the decision made is not correct.
3. It gives less correct answer than the 1% level of significance.	3. It gives more correct answer than the 5% level of significance.
4. This is useful for rejecting or accepting the null hypothesis at 5% level.	4. This is useful for the rejecting or accepting the null hypothesis at 1% level.

Differentiate between 't' and 'F' Test

't' test	'F' test
1. To test the sample mean with population mean.	1. To test the equality of the population variances.
2. To test the difference between two mean	2. To test multiple regression coefficient.
3. To test correlation & regression coefficient i.e. rbyx & bxy.	3. To test linearity of regression.
4. To test partial correlation coefficient.	4. To test equality of means.

Differentiate between Skewness and Kurtosis

Skewness	*Kurtosis*
1. When the mean of the normal distribution is X & standard deviation σ & other constants as follows; $\mu_2 = \sigma^2$, $\mu_3 = 0$, & $\mu_4 = 3\sigma^4\sigma_1$ or moment of coefficient of Skewness = $\mu_3^2 / \mu_2^3 = 0$	1. B or moment coefficient of kurtosis $\sigma^4 / \sigma_2^2 = 3\sigma^4 / \sigma^4 = 3$
2. Hence normal distribution is perfectly symmetrical.	2. Hence the normal distribution is not perfectly symmetrical.
3. When β_1 is equal to zero then & then only normal distribution is symmetrical.	3. If the value of β_2 is more than 3, the curve is leptokurtic & if the value of β_2 is less than 3 the curve is platykurtic.

Differentiate between Binomial and Normal Distribution

Binomial distribution	*Normal distribution*
1. Binomial distribution is a discrete type of distribution.	1. Normal distribution is a continuous distribution.
2. The mean = np & the standard deviation = σ = $\sqrt{npq}$	2. The mean, mode, median are same in value.
3. The area under the curve is not equal to the zero.	3. The area under the curve is equal to unity.
4. The binomial distribution can be change to normal distribution when 'n' is very large.	4. The normal distribution does not change in to binomial distribution.

Differentiate between Variance and Coefficient of Variation

Variance	Coefficient of variation
1. The square of the standard deviation is known as the variance	1. The ratio between the standard deviation to the arithmetic mean in percentage is known as C.V.
2. Variance is expressed in term of units in which the data or item measured.	2. The coefficient of variation is the unit less quantity & it is express in percentage.
3. Formula $\sigma = \Sigma(xi - x)^2/n$	3. Formula C.V. $= \sigma /x \times 100$

4

OBJECTIVE TYPE QUESTIONS

SET-A: FILL IN THE BLANKS

S. No.	Statement

1. In an ideal classification number of classes should be between the ranges **5 to 15**.

2. Height of rectangular in a histogram is proportional to **Frequencies**.

3. Characters can be classified as **quantitative** and **qualitative**.

4. A quantity that varies from individual to individual is called **continuous variable**.

5. Number of leaves per plant is **discrete** variable whereas height of the plant is **continuous** variable.

6. The classification of flower according to their colour is **qualitative** type of classification.

7. **Arithmetic** mean is a best measure of central tendency.

8. **Arithmetic mean, median, mode, geometric mean, harmonic mean, weighted mean** are there important measures of central tendency.

9. Median of the series 3, 6, 4, 7 and 1 is 4. i.e. [N+1÷ 2].

10. The observation, which divides the given data into two equal parts, is known as **median**.

11. The class having maximum frequency is known as **modal class**.

12. The median and mean are 35 and 30 respectively than value of mode is **45**, [Mole = 3 median – 2 mean].

13. The value of the variable of which the frequency curve reaches maximum value is **mode value**.

14. In **normal or symmetrical** distribution values of mean, median and mode are same.

15. The value of **mode** and **median** can be determined graphically.

16. **Median** and **mode** are positional averages where as arithmetic mean is a computational average.

17. The algebraic sum of deviations taken from arithmetic mean is always equal to **zero**.

18. The algebraic sum of squares of deviation taken from arithmetic mean is always **minimum** or **least**.

19. **Range** is the difference between the value of the smallest item and the value of the largest item in the distribution.

20. **Standard deviation** is best measure of dispersion.

21. **Standard deviation** is absolute measure of dispersion where as **coefficient of variation** is a relative measure of dispersion.

22. Square of standard deviation is called **variance.**

23. Square root of variance is called **standard deviation.**

24. Coefficient of variation = **Standard deviation + Arithmetic mean × 100.**

25. The value of correlation coefficient lies between **+1 and –1.**

26 In case of perfect positive correlation the value of **r = 1.**

27. In case of perfect negative correlation the value of **r = –1.**

28. When there is no relation between two characters the value of **r = zero.**

29. For qualitative characters we use **Rank correlation** method of correlation.

30. For quantitative characters we use **Karl Pearson's** method of correlation.

31. The correlation between high of the experimenter and grain yield of the crop of the experiment, which he has conducted, is **zero** correlation.

32. By **scatter diagram** method of correlation we can not get an idea about the degree of correlation but we can only say whether there exists correlation between two characters or not.

33. **Karl Pearson's** correlation method gives us degree as well as direction of correlation between two characters.

34. **Karl Pearson's** and **Spearmans rank correlations** are two methods for study of correlation coefficient between two characters.

35. Correlation coefficient is **remaining same or unaffected** by change of origin and scale.

36. Square of the correlation coefficient is called **coefficient of determination.**

37. The statistical tool with the help of which we are in a position to estimate the unknown values of one variable from values of another is called **Regression.**

38. In study of rainfall and yield, **rainfall** is independent variable whereas **yield** is dependent variable.

39. The straight line obtained by minimizing the sum of squares of deviations parallel to Y axis is called **Regression line of Y on X, [Y = a+bX].**

40. The straight line obtained by minimizing the sum of squares of deviations parallel to X axis is called **Regression line of X on Y, [X = a+by].**

41. The two lines of regressions viz. line of regression of Y on X and line of regression of X on Y intersects at the point a **[of their mean values].**

42. A value gives the intercept of the line of regression with Y-axis.

43. If Z is a normal variate then p $\{-1(-2\ y-1)\}$ = ...

44. Mean of the Binomial distribution = **np and α = √npq**

45. If **n = 25** and **s = 3.6** then standard error of mean = 0.72.

46. In an rxc contingent table, the degrees of freedom of X^2 =......

47. Two tailed test of significance is applied if Ho $\mu = \mu_o$ and H1:...

48. If **x = 15, μ = 10** and **σ = 2.5** then the standard normal variate Z = 2.

49. If b_{yx} = 0.48, b_{xy} = 1.330 then **r = 0.77**

50. If Σ_y^2 = 7600 and Σ_{xy} = 1900 then the coefficient of determination = **475**

51. $P(Z) = P(-0.0 \le r, Z \le 0) = \ldots\ldots\ldots$

52. If $C_{11} = 0.0008$ and M.S.S. due to deviation from $sb_1 = S\sqrt{C_{11}}$

53. If A [8 2] then A^{-1} **[1/8 1/2] [10 5] [1/10 1/5]**

54. If the equation of plane of regression is $Y = a + b_1 X_1 + b_2 X_2$ then **(i) $Y = a + b_1 X_1$** (ii) $Y = a + b_2 X_2$ (iii) $X_1 X_2 = a + b_1 b_2 Y$

55. The mean incomes of 75 male labors is Rs. 175 per month and mean income of 50 female labors is Rs. 150 per month then the mean income of the whole group of labor is Rs. **165** per month.

56. If $X = 15$, $\mu = 10$, $s = 2.5$ then the standard normal value of Z = **2.**

57. $byx = 0.4$, $bxy = 1.2$ then $r_{xy} = 0.69$

58. Representative part of the population is called **sample.**

59. Any summarizing quantity calculated from the population is called **population parameter.**

60. Any summarizing quantity calculated from the sample is called **sample parameter** [statistics].

61. ……………… is an estimate of population parameter.

62. Probability of an event = p = number of success of an event ÷ total number of trials.

63. Mean of binomial distribution is = np

64. Standard deviation of binomial distribution = $\sigma = \sqrt{npq}$

65. Mean and variance of Poisson distribution = l, $\sigma = \sqrt{l}$

66. Binomial and Poisson are discrete probability distributions whereas Normal is continuous probability distributions.

67. For normal distribution mean, median and mode are same.

68. The standard normal variate has mean zero and variance one.

69. The normal distribution has mean r and variance = δ

70. Total area under probability curve and X axis is considered to be equal to one.

71. $\int$ and δ are parameters of normal distribution.

72. Standard deviation of sample means is called ————

73. Null hypothesis is a statement about **equality.**

74. If sample size is greater than **30** we call it as large sample

75. If the sample size less than 30 we call it as **small** sample.

76. To compare two population means we use '**t' test.**

77. To compare two population variances we use '**f' test**

78. To compare more than two means we use '**t' test**

79. To test association between two characters we use x^2 **test** (chi-square test)

80. If calculated value is greater than table value we **reject** the null hypothesis

81. If calculated value is less than table value we **accept** the null hypothesis

SET-B: STATE TRUE OR FALSE

1. The midpoint of the class is known as class interval.

2. The value of the median is 11. If two new items having values 4, 13 are added to the series then value of median will remain same.

3. The difference between lower and upper limit of class is called as range.

4. For characters which can be measured by some units we use qualitative classification.

5. Height of plant is a continues variable.

6. Number of leaves is a discrete variable.

7. Height of histogram is proportional to the frequency of the class.

8. We can obtain value of mode by Histogram.

9. We can obtain value of median by frequency curve.

10. Arithmetic mean can be obtained by graphical method.

11. The value of arithmetic mean is not affected by fluctuation of value of extreme items.

12. Algebraic sum of squares of deviation taken from arithmetic mean is always zero.

13. Algebraic sum of squares of deviation taken from median is always minimum.

14. Arithmetic mean is the value which divides the data into two equal parts.

15. Standard deviation is square of variance.

16. Standard deviation is a best measure of central tendency.

17. Standard deviation is useful for comparison of two series.

18. Standard deviation is relative measure of dispersion whereas coefficient of variation is absolute measure of dispersion.

19. In calculation of mean deviation we ignore algebraic signs.

20. Value of standard deviation consists of coefficient of variation there is no unit.

21. Standard deviation of means is called standard error of mean.

22. We prepare a character having maximum value of standard deviation.

23. Small value of standard deviation indicates that there is consistency in the values of character.

24. Series having values 5, 5, 5, 5, 5 will have standard deviation equal to 5.

25. Value of standard error increases as sample size increases.

26. The value of simple correlation coefficient lies between range 0 and 1.

27. If value of correlation coefficient is 1.25 we say that there is no correlation between the two characters as the value is not satisfying the range of correlation coefficient.

28. Scatter diagram gives us the degree of correlation between two variables.

29. Karl Pearson's correlation coefficient is unaffected by change of origin and scale.

30. Correlation coefficient is symmetric in X and Y i.e. $r_{xy} = r_{yx}$.

31. Regression coefficient is symmetric in X and Y i.e. $b_{xy} = b_{yx}$.

32. Value of regression coefficient lies between the range -1 and +1.

33. If value of $b_{yx} = 0$, the line of regression of Y on X lies parallel to X axis.

34. If a = 0 the line of regression passes through origin.

35. Value of regression coefficient is unaffected by change of origin and scale.

36. Binomial distribution is continuous probability distribution.

37. For binomial distribution mean, median and mode is same.

38. Normal distribution is symmetric about mean.

39. Standard normal distribution is symmetric about X = 0.

40. $P (0 < X < \infty) = 1$.

41. F test is used for comparisons of two variables.

42. 't' is used for testing association between two attributes.

43. X^2 test is used to compare two variables.

44. Null hypothesis is the statement about population parameters.

45. Null hypothesis is the statement about statistics.

46. When calculated value is less than table value we accept null hypothesis.

47. Paired 't' test is used to compare two means of independent samples.

48. Z test is used when sample size is less than 30.

49. To increase the precision of conclusion we should decrease the value of level of significance.

50. To increase the precision of conclusion we should increase the value of level of significance.

51. Arithmetic mean is a best measure of dispersion.

52. A river has mean depth of water of two feet so that any adult can go through it very easily without swimming.

53. $-1 \geq "r_{xy} \geq" +1, -1 \leq" b_{yx} \leq" +1.$

54. $-1\leq" r_{yx} \leq"+1, - " \leq " b_{yx} + \leq" +".$

55. The median weight of fifteen students is 40 kg four move students of weight 43 kg, 39 kg, 46 kg and 38 kg are added to the group. Then value of median and mean will remain same for the whole group.

56. If the sum of squares of deviation parallel to X axis is minimized then the line of best fit is called as the line of regression of X on Y.

57. Value of multiple correlation coefficient lines between 0 and 1 whereas partial correlation coefficient lies between -1 and +1.

58. Null hypothesis is a statement of equality which can be possibly disproved.

59. If the difference sample mean and population mean is found to be significant the sample does not belong to the population.

60. Coefficient of correlation is affected by change of origin but not by change of scale.

61. Coefficient of regression is affected by change of origin and scale.

62. Degree of freedom of a statistics is the number of independent comparisons on which it is based.

63. Degree of freedom of a statistics is always equal number of observations minus one.

64. 'F' test is used for testing the significance of difference between two means while 't' test is used for testing the ratio of two variances.

65. Binomial distribution is a continuous probability distribution.

66. Correlation measures the relationship between the variations of the two variables.

67. Normal distribution is a discrete probability distribution.

68. In case of are tailed test 5 per cent table value of Z is equal to 1.96.

69. Coefficient of correlation indicates the degree and direction of the relationship between the variations of two variables.

70. $-1\leq" R1 (23) \leq" +1.$

71. If $b_{yx} = -0.4$ and $r = 0.6$ then $b_{yx} = + 0.9.$

72. Correlation coefficient is the geometric mean between the regression coefficients.

73. If one of the regression coefficients is greater than unity, the other must be less than unity.

74. (X, Y) lies on both lines of regressions.

75. For testing the significance of difference between three sample means we use 't' test.

76. Line of regression of Y on X is given by $(X-X) = b_{yx} (X-X)$.

77. Standard deviation is a measure of variation used for comparing the variation between two series.

78. Coefficient of determination is the square of coefficient of correlation.

79. Coefficient of determination = Explained variation ÷ total variation.

80. Algebraic sum of deviations of all the variate from the median is always equal to zero.

SET-B: ANSWERS KEY

1.	False	21.	False	41.	True	61.	False
2.	True	22.	False	42.	False	62.	True
3.	False	23.	False	43.	False	63.	True
4.	False	24.	False	44.	False	64.	False
5.	True	25.	False	45.	False	65.	False
6.	True	26.	False	46.	True	66.	False
7.	True	27.	False	47.	True	67.	False
8.	True	28.	False	48.	False	68.	False
9.	True	29.	True	49.	False	69.	True
10.	False	30.	True	50.	False	70.	False
11.	False	31.	False	51.	False	71.	False
12.	False	32.	True	52.	False	72.	True
13.	False	33.	True	53.	False	73.	False
14.	False	34.	True	54.	False	74.	False
15.	False	35.	False	55.	True	75.	False
16.	False	36.	False	56.	False	76.	False
17.	False	37.	False	57.	False	77.	False
18.	False	38.	True	58.	True	78.	True
19.	True	39.	False	59.	False	79.	False
20.	True	40.	False	60.	False	80.	False

REFERENCES

Cavalli, L.L. 1952. **Analysis of linkage in quantitative inheritance.** In: E.C.R. Reeve and C.H. Waddington (Eds). Quantitative Inheritance, H.M.SO. London, pp. 135-144.

Gupta, S.P. 1976. **Statistical Methods.** Sultan Chand & Sons Publisher, 4742/23, Daryaganj, New Delhi.

Mather, K. and Jinks, J.L. 1971. **Biometrical Genetics,** 2nd Edn. Chapman and Hall Ltd. London.

Panse, V.S. and Sukhatme, P.V. 1967. **Statistical Methods for Agricultural Workers.** Indian Council of Agricultural Research, New Delhi. pp. 347.

www.ingramcontent.com/pod-product-compliance
Lightning Source LLC
Chambersburg PA
CBHW021436180326
41458CB00001B/293